Photo. by Holmes, N.Y.

Endicott & Co. l

Wells of the Mc Kinley Oil Company of New York.

VIEW ON OIL CREEK, PENNSYLVANIA.

A PRACTICAL TREATISE ON COAL, PETROLEUM, AND OTHER DISTILLED OILS,

By ABRAHAM GESNER, M.D., F.G.S.

Second Edition, Revised and Enlarged.

By GEORGE WELTDEN GESNER,
CONSULTING CHEMIST AND ENGINEER.

Stanton and Williams Wells, Owned by Fountain Petroleum Co.

NEW YORK:
BAILLIÈRE BROTHERS, 520 BROADWAY.

LONDON:
H BAILLIERE,
219 REGENT ST.

MELBOURNE:
F. BAILLIERE,
COLLINS ST.

PARIS:
J. B. BAILLIERE ET FILS,
RUE HAUTEFEUILLE.

MADRID:
C. BAILLY-BAILLIERE,
CALLE DEL PRINCIPE.

1865.

R. CRAIGHEAD, PRINTER,

81, 83, *and* 85 *Centre Street, New York*

PREFACE.

THE work before the reader has been prepared with a desire to aid the manufacturer of distilled oils in his avocation, and to record the most important facts regarding the various raw materials used by him. To assist the engineer and machinist, drawings, taken from the original plans used by the author in the erection of petroleum and coal-oil apparatus, have been inserted. Original drawings of petroleum wells and boring tools have also been added, to convey a correct idea of the mode of procuring one of the chief articles used by the oil manufacturer. A great deal of valuable information relating to the production of hydrocarbon oils is scattered throughout the French, German, and English scientific journals, and is contained in numerous patents to which the public have not convenient access. The information collected from these sources that seemed applicable to the present treatise, has been carefully recorded. The author of the first edition of the work left, at the time of his decease, many memoranda connected with it, intending to insert them in a second edition. The author of the present edition has endeavored to carry out his father's intentions regarding it, and has made such revisions and enlargements as seemed called for.

Care has been taken to present such facts regarding petroleum wells and petroleum as may be useful to those interested in them.

It is hoped that the professional chemist may find in this treatise something to interest him, and especially as regards the homologous compounds of carbon and hydrogen, and the manipulation of aniline.

59 WILLIAM STREET, NEW YORK, *July*, 1865.

CONTENTS.

CHAPTER VI.

CHAPTER VII.

CHAPTER VIII.

COAL, PETROLEUM, AND OTHER DISTILLED OILS.

CHAPTER I.

Early records and progress of the distillation of oils from coals and other bituminous substances.—Introduction of kerosene, patents, petroleum, varieties of coals; their origin and composition.—Effects produced upon coals by heat.—Variety of oils distilled from petroleum, coals, bitumen, etc.—Products of common bituminous coals, etc.

In a treatise devoted to the manufacture and purification of oils, it might be deemed proper to consider the oleaginous substances derived from the animal and vegetable kingdoms; but the present treatise is intended to give a descriptive account of the mineral oils only, and the modes by which they are manufactured and purified.

The rapid advances made during the last fifteen years in developing mineral oils, and their growing importance to the world for illuminating, and various other purposes, give them a value not rivalled by many articles of commerce.

The ancient inhabitants of different parts of the world, civilized and barbarous, were acquainted with those natural oils which flow from the earth, namely, mineral oil, or naphtha, bitumen, etc. Herodotus, the Greek historian, who composed his history 440 years before Christ, speaks of a place called Arderrica, thirty-five miles from Susa, where there were wells which yielded bitumen, salt, and oil. In place of a bucket, half of a wineskin was used to draw the product of these wells, which was permitted to settle in tanks, in which the bitumen and salt collected and hardened, the oil being drawn off into casks. The oil was called by the Persians "Rhadinace," was black, and had an unpleasant odor.

The Persians, Burmese, and other nations still continue to employ those substances in their crude state to give light, and for medicinal purposes. As early as 1694 Eeele, Hancock, and Portlock made "*pitch, tar, and oyle out of a kind of stone,*" and obtained patents therefor. In 1761 oils were distilled from black bituminous shale, and were employed in the cure of certain diseases, as stated in Lewis's Materia Medica for that year.

More than a century ago oils were obtained by the distillation of coals, but the purification of those oils, and their application to the common requirements of life, have been slow in their progress, and are not even now brought to perfection. The papers of the Royal Society of London, the Philosophical Transactions, and other European publications, give accounts of the distillation of oils from coals and other bituminous substances. In 1781 the Earl of Dundonald obtained oils from coals by submitting them to dry distillation in coke ovens, like those employed by some manufacturers of the present day for the same purpose. Laurent, Reichenbach, and others distilled the tars obtained from bituminous schists. These tars were purified in some degree by Selligue, and the oils subsequently obtained an extensive sale in Europe for burning in lamps, and for lubricating machinery. Many other chemists have from time to time contributed improvements in the purification of hydrocarbon oils.

The discovery of coal gas brought a new class of oils to

the notice of the chemist, but the purification of those oils, and their application to useful purposes, have been but slowly advanced.

The first successful attempt to manufacture oils from coals in America was made by Dr. Abraham Gesner. Oil from coal was made and consumed in lamps by him in his public lectures at Prince Edward's Island, in August, 1846, and subsequently at Halifax, Nova Scotia, accounts of which are still extant. The patents afterwards obtained for his improvements, known as the "Kerosene Patents," were sold to the North American Kerosene Gas Light Company, and the oils were manufactured and sold under the denomination of "Kerosene Oil."* Several patents, obtained by other persons at later dates, are but modifications of the modes of manufacture previously laid down, and contain but little that is new in principle.

The retorts, stills, and other appliances for this kind of manufacture have been constantly varied, and have not yet been so perfected as to meet the general approval of manufacturers. In a future chapter these varieties of apparatus will be described.

Patents were granted in England, in 1847, to Charles B. Mansfield, for "an improvement in the manufacture and purification of spirituous substances and oils applicable to the purposes of artificial light," etc. Mr. Mansfield's operations appear to have been chiefly directed to the coal tar of gas works, from which he obtained benzole. He was perhaps the first to introduce the benzole or atmospheric light, which is described at length in his specifications.

James Young, of Manchester, secured a patent in England (Oct. 7, 1850), and subsequently in the United States (March 23, 1852), for "the obtaining of paraffine oil, or an

* From κηρός, wax, and ἔλαιον, oil.

oil containing paraffine, and paraffine from bituminous coals." These patents have been the subject of much discussion, and law proceedings have been instituted by Mr. Young against several oil companies for infringements of his alleged rights.

The manufacture, nevertheless, extended very rapidly to the chief cities of the Atlantic seaboard, and of the coal districts of the interior. The great cheapness of the oil procured by the distillation of petroleum has, however, almost caused the coal distillation to be suspended. It will only be resumed when the petroleum wells cease to yield sufficient oil for the various purposes to which it is now applied. From a calculation made in 1861, it was shown that whenever crude petroleum reached an average price of thirty-five cents per gallon in the American markets, the coal oil distiller could afford to resume business. It is not at all probable that any claims of patentees which seek to monopolize the distillation of coal for oil making, will be enforced, should the time again arrive when such distillation can be profitably carried on.

The progress of discovery in this case has been gradual. It has been carried on by the labors, not of one mind, but of many, so as to render it difficult to discover to whom the greatest credit is due. It is, notwithstanding, just to admit, that from the facts disclosed in the before-mentioned patents, a spirit of inquiry was aroused, and experiments were multiplied.

The introduction into common use in America of oils distilled from coal, bitumen, and incidentally petroleum, was accomplished by the North American Kerosene Gas Light Company of New York, in the early part of 1854. This Company was formed to work under the Kerosene Patents, and Messrs. John H. Austen and George W. Austen, the agents of the Company, first sold the oil produced by the patentee at the Company's works on Newtown Creek, Long Island, N. Y., under the name of "kerosene." The Messrs. Austen found great difficulty in selling the oil. The refining process was not so well understood at that time as at present, and the odor was not agreeable. The beauty of the light obtained from it, however, was sufficient to gradually overcome the objection on the score of odor. Its supposed explosiveness was also urged against it by those interested in the camphene and burning fluids. These two latter are very explosive compounds, but that fact was overlooked by the opponents of kerosene, which, as it was originally manufactured by Dr. Gesner, was quite as safe an article as ordinary whale oil.

There was a serious drawback to the general use of kerosene in the want of a cheap and proper lamp in which to burn it. To John H. Austen, Esq., is due the credit of supplying this very necessary adjunct to the coal-oil business. This gentleman found in Vienna a burner which was suited to light hydrocarbon oils, and under the name of the "Vienna burner" brought it into use. This burner has formed the model upon which great numbers of patents have been framed. It has not always been improved by inventors.

With a view to a proper arrangement of the subject, various materials capable of yielding oil by distillation will be considered in regular order.

The chief of these are petroleum, bitumen or asphaltum, coals, bituminous shales, sands and clays, peat, caoutchouc, gutta percha, and the tars produced in the manufacture of stearine.

When organic bodies are exposed to heat, with the free admission of air, they undergo combustion. The greater part of the carbon is expelled in smoke, or in carbonic acid, the hydrogen in water, or carburetted hydrogen, and the nitrogen, if any be present, escapes in some compound of ammonia; but if those substances have heat applied to them in close vessels, there are new results, and a greater interchange of elements takes place.

In 1730 Hales distilled substances to discover if they contained air. In 1773 the same gentleman and Dr. Watson obtained gas from coals, and in 1786 Lord Dundonald burned the gas that arose from his coke ovens at the ends of iron pipes for the amusement of his friends. In 1792 Mr. Murdoch commenced lighting buildings with coal gas, and since that period gas lighting has been extended to every quarter of the globe. Besides the gas employed for illumination, it was thus early observed, that other gases and oils were produced by the distillation of coals. The discovery of coal oils is therefore as old, if not older, than the discovery of coal gas, and cannot now be justly claimed by any living man.

The discovery of coal oils has led, no doubt, to the discovery of the value of petroleum, and those bituminous substances most resembling it.

When substances composed of carbon, hydrogen, and oxygen are submitted to dry distillation, the first effect is to remove oxygen from the body in the form of carbonic acid, or water. After the oxygen has been removed carbon and hydrogen escape, as carburetted hydrogen, or olefiant gas. If some of the acids are distilled they lose oxygen in the form of carbonic acid and water, and are converted into new acids. Organic acids distilled with strong bases part with the elements of carbonic acid, which uniting with the base and the acid, minus the carbonic acid, comes over in the form of a new product.

If a quantity of coals be placed in a suitable retort, with a condensing apparatus attached, and heat be gently and gradually applied thereto, the first result will be the escape of water in the form of vapor, or steam, and frequently mixed with an extremely light, volatile, and inflammable hydro-carbon, which is but partially condensable into a spirit, or oil. The hygroscopic water contained in the coal appears in the form of vapor, sometimes mixed with carbonic acid, and if the coal contained nitrogen ammonia is among the products. Then as the heat is increased a series of oils of different specific gravities are condensed, the lightest or first distilled having the character of a spirit rather than an oil; finally, when the heat has been raised to 750° or 800° Fah., gas, free carbon, and a number of pyrogenous substances appear, known as *dead oil*, which mixes mechanically with the aqueous products, and settles to the bottom of the receiving vessel. Throughout the distillation more or less water, formed by the oxygen and hydrogen present, continues to flow. Usually in proper retorts the oils will all distil over at a temperature of 750° Fah. A higher degree of heat produces permanent gases from any volatile matter that may remain in the charge; but besides the elements before-mentioned, coal frequently contains sulphur and other minerals, which, by entering into new combinations during the distillation, yield other compounds, the *modus operandi* of whose formation is not well understood. In the retort there remains a quantity of fixed carbon, or coke, united to the ash, which usually consists of silica, alumina, lime, and the oxides of manganese and iron.

The results here described are greatly modified by the kind of coals used, the degree of heat applied, and the mode by which the oleaginous vapors are condensed. The shape of the retort, the weight, or thickness of the charge, and the position and size of the discharge-pipe, also have an influence over the yield of oil, and the time required for its production.

In general, coals which yield the greatest amount of volatile matters, exclusive of hygroscopic moisture, afford the most oils, and estimates are often formed of their value by a simple test of the weight of matter expelled by the application of a moderate degree of heat. The test, however, is often delusive, as some coals expel much more free carbon during the distillation than others, and the sulphur contained in coal adds nothing to the oil, while it constitutes a part of its volatile products. Nor does such a test afford much information regarding the quality of the oils a given quantity of coal will supply. A long smoky flame is indicative of much free carbon, a shorter and more luminous flame denotes that there will be much fixed carbon in the coke. Some varieties of coals are peculiarly adapted to the manufacture of gas, as their chief products by heat are carburetted and bicarburetted hydrogen; such coals do not always contribute the most oils.

It is of the utmost consequence to the manufacturer of coal or petroleum oils to know the quality as well as the quantity of the oils any one material will afford. For this the only reliable test is to submit the material to dry distillation, and the whole process by which the oils are purified.

It will be seen hereafter that coals, coal shales, asphaltums, petroleums, and other bituminous substances, yield not one, two, or three oils; but series of homologous compounds. Some members of these series are of high specific gravity, some of low, or, as the oils are called, heavy and light; the light being eupion, or benzole, the heavy the oil pressed from paraffin, and, finally, the solid, as paraffin, naphthalin, etc.

These several series of hydrocarbons are greatly influenced by the heat employed in their distillations, the condensers, and, finally, their mode of treatment. Again, there are not two kinds of coal that will give the same products, even by the same modes of manufacture. Some yield much light, others much heavy oil; some send over much paraffin, and what are called by manufacturers impurities, namely, naphthalin, carbolic acid, copnomor, etc., ever attending the distillates.

Few common bituminous coals can be successfully employed in the oil manufactory; their distillates abound in creosote, or carbolic acid, and their purification is expensive. The modes of refining the oils from petroleum, coal, bitumen, and other bituminous substances, will be given in their proper places. For the present the author will confine himself as much as possible to the description of the various substances capable of yielding oil by distillation, beginning with petroleum.

CHAPTER II.

Petroleum of the United States.—Theories as to its Origin.—Geological Features of the Petroleum Regions.—Petroleum Wells.—Boring for Petroleum.—Boring Tools and Machinery.—Petroleum of Canada, South America, Trinidad, Barbadoes, Burmah, etc.

ALTHOUGH petroleum has been known to exist in various parts of the world for centuries, it was not until the oils derived from the distillation of coal and bitumen had been brought into use, that the attention of the business world was attracted by it. Now that its value is becoming appreciated, the petroleum-producing portions of the globe are being rapidly explored, and it is not at all improbable that discoveries even more wonderful than those of the past seven years may reward the efforts of those who are venturing time and means in searching for this very important hydrocarbon.

The petroleum wells of the United States claim, from their number and productiveness at the present time, the chief place in a work devoted to the history and chemical treatment of petroleum and coal oils.

Petroleum has long been known to exist in the State of New York. Under the name of "Seneca Oil" which it derived from an Indian tribe, petroleum was formerly collected in Chautauque County, N. Y., and in Crawford County, Pennsylvania, and sold for medicinal purposes. It is not improbable that the country lying beyond the Alleghanies in New York, may yet be found rich in petroleum. Pennsylvania is the largest oil producing state. Ohio is also yielding largely. West Virginia has produced a large quantity. Kentucky bids fair to equal her sister states in the petroleum production. Tennessee, Georgia, Alabama, Missouri, and Texas, are known to contain petroleum springs. Sour Pond, so called from the circumstance that during part of the year the waters are acid to the taste, is a pitch lake resembling that of Trinidad. It is between Beaumont and Liberty in the latter state.

Arkansas and Missouri are rich in bituminous sands, shales, and clays. Petroleum has been found in Illinois, Indiana, and Michigan.

From recent explorations, it would seem that California is capable of yielding an enormous quantity of petroleum. The existence of asphaltum and semi-solid bitumen at Santa Barbara in Southern California, had been known since 1792, but no generally published account of the extent of the deposit was had until its survey by Professor Silliman in 1864.

It has afforded much interest to the geologists and chemists of the day, to ascertain from the geology of the petroleum districts, the origin of petroleum itself.

Were the petroleum now produced by the various wells of the same quality, and the strata from which they are derived of the same character, the obstacles in the way of reasoning out a theory of their origin would be very much removed. But even the petroleums of the United States alone differ materially, as will be hereafter seen.

The theory that the petroleum of Canada, which occurs in the older Silurian rocks, is derived from the decomposition of vast numbers of marine animals, is not an unreasonable one. In distillation the Canada petroleums yield acroleine, an oil which is obtained from animal oils and fats. The vapor of acroleine is very pungent, and attacks the mucous membrane of the throat and lungs, causing great irritation. Fish oils yield it by distillation. It is not found in the petroleums of Pennsylvania.

It is remarked by a learned writer of the day, that "the transformation of woody fibre into oil is a chemical change, taking place always out of contact with atmospheric air and usually under water, but by no means necessarily connected with any particular geological period, as, for example, the coal epoch, with which many intelligent people associate it."

During the passage of vegetable substances into coal, there is an escape of vast quantities of carbon combined with hydrogen. It is only necessary that the gases of these elements should be condensed to produce hydrocarbon oils. The operation is a decomposing and a combining one, and the new combinations formed during the transmutation of wood into coal, have a close analogy to those produced during the distillation of wood without the admission of air. The gases generated in strata of coal and coal strata, are always under great pressure, which tends to their condensation and consequent formation of oil.

That coal has been derived from vegetables is undoubted. Peat and wood are found to pass by insensible shades into lignite, lignite into compact bituminous coal, and the end of the transformation appears in the anthracite, from which nearly all the hydrogen has been expelled and carbon remains.

From the expulsion of oxygen, carbon, and hydrogen from wood, and the variety which it presents until it forms true coal, heat has not been absolutely necessary, although it has doubtless exercised a powerful influence in connexion with those chemical changes ever going forward in the earth.

The condensation of carbon and hydrogen producing oil,

and the fact of strata of coal and shale before they reach the maximum of carbonization giving out these elements in great quantities under pressure, and the tendency of these gases and oils to diffuse themselves, are fair reasons for finding the oil in formations bearing no traces of vegetables.

Many theories have been advanced regarding the origin of petroleum. By one writer it is supposed that "the petroleum of Pennsylvania arises from the distillation by subterranean heat of the hydrocarbon agents resident in the carbonaceous strata underlying the oil region." By another, "that the great beds of anthracite coal of Pennsylvania, on the southerly slope of the Alleghanies, are merely the residuary coke, as it were, of a distilling process, which has converted their bituminous matter into oil, and distributed it by some convulsion of the earth through the formation beyond the mountain range."

So far, however, as research has gone, it has failed to present a theory acceptable to all. It must be agreed, however, that all petroleum could not have had precisely the same origin, and that a theory which might fairly apply to the origin of the petroleum of one district, would not at all apply to that of another.

In Pennsylvania and Ohio the petroleum is found in the Devonian formation. In Canada, it is found in the Silurian limestone. In Pennsylvania, alternate beds of the Utica slate and sandstone are pierced by the boring tools. The evidences of great disturbance of the strata are numerous. In West Virginia, they are seen in the anticlinal ridges which traverse the petroleum region. At the base of these ridges, and where the surface indicates the greatest disturbance beneath, the petroleum wells are usually located.

In the sandstone beds there are large cavities filled with petroleum, gas, and water, and when these are reached by the boring tools, there is a great and violent escape of their contents until they have become exhausted. It sometimes happens that two wells near together will enter the same crevice, and either the sooner exhaust the supply of petroleum, or drain it away, the one from the other.

As the formation of petroleum in the earth is a slow process, it is not probable that the supply can keep pace with the demand made upon it where the wells are sent down too near together. None of the wells have continued to flow for any great length of time.

PETROLEUM WELLS.

The most productive are situated in the northwestern part of Pennsylvania, in Venango, Crawford, Clarion, and adjoining counties. A large area of country between Pittsburg and Lake Erie, and that portion of it drained by the Alleghany River and its tributaries, Oil Creek, French Creek, Tionesta Creek, and smaller streams, has been found to yield petroleum. Beaver County, Pennsylvania, near the Ohio River, has also become known as an oil-producing region. The locality known as Smith's Ferry has been successfully explored for oil. In West Virginia the counties of Pleasant, Ritchie, Wood, and Wirt, comprise the oil region. They are near or bordering upon the Ohio River, and are drained by the Little Kanawha and Goose, French, Bull, Horseneck, Calf, and Stilwell Creeks. The Ohio petroleum district is principally in Washington, Noble, and Morgan Counties, on the Great and Little Muskingum, and on various other streams, such as Duck Creek, Paupaw, Wolf, and Federal Creeks. In Indiana petroleum is found on Little Blue River, in Crawford County. In the other States before mentioned, petroleum wells are in progress. The California and Kentucky petroleum regions will no doubt soon add greatly to the supply.

The introduction of petroleum into market took place about three years after the oils obtained from coal had been in use. Professor Silliman analysed a sample in 1854. An oil for lamps, called Carbon oil, was introduced in the New York market, by A. C. Ferris, Esq., in 1857. It was at first sold after being distilled, but not treated, but was afterwards refined. It was procured from an old salt well at Tarentum, on the Alleghany, not far from Pittsburg, where it was found in such quantities as seriously to interfere with the salt manufacture.

The boring of petroleum wells was begun by Mr. E. L. Drake, who began a well at Titusville, on Oil Creek, Pennsylvania, in 1858. Mr. Drake was at first not successful in finding the petroleum at the depth which it was supposed was sufficient, but persevered, and at last struck a fine vein of petroleum.

Since that time the business of procuring it has become one of the most extensive in America. More than four hundred Petroleum Companies have been formed. Many of these companies may not succeed in their efforts to obtain oil, and many of them may never practically set about the business of sinking oil wells, but enough will remain to develop thoroughly the oil region of the United States.*

The estimated yield of petroleum at the present time by

---

* Remarkable success has in some instances rewarded the enterprise of Petroleum Companies. Among them may be named the McKinley Oil Company of New York. This Company paid twenty-two per cent. in dividends upon a capital stock of $250,000 within a period of four months. The frontispiece to this work represents a view of a portion of the Company's property.

the oil wells of the United States is 6000 barrels, or 240,000 gallons per day. Though certain wells have for a short period yielded 1000 barrels or 40,000 gallons per day, and even more than that quantity, the average yield of all will not be much over 5 barrels, or 200 gallons per day. An oil well yielding 100 barrels, or 4,000 gallons per day, is not a common thing to find in the oil region; half that quantity would be considered a very good yield.

THE EXPORT OF PETROLEUM.

The following statement shows the export of petroleum from the United States to all parts of the world:

From New York to	1864. Gallons.	1863. Gallons.
Liverpool,	734,755	2,156,851
London,	1,430,710	2,576,331
Glasgow, etc.	368,402	414,943
Bristol,	29,134	71,912
Falmouth, E.,	316,402	623,176
Grangemouth, E.,	——	425,334
Cork, etc.,	8,310,362	1,532,257
Bowling, E.,	87,164	——
Havre,	2,324,017	1,774,890
Marseilles,	1,982,075	1,167,893
Cette,	4,800	——
Dunkirk,	232,803	——
Dieppe,	79,581	46,000
Rouen,	——	143,646
Antwerp,	4,149,821	2,692,974
Bremen,	971,905	903,004
Amsterdam,	77,041	436
Hamburgh,	1,186,080	1,446,155
Rotterdam,	532,926	757,240
Gottenburg,	38,813	——
Cronstadt,	400,376	88,069
Cadiz and Malaga,	55,674	33,284

Tarragona and Alicante, .	16,823	33,000
Barcelona, ...	25,500	———
Gibraltar, ...	69,181	308,450
Oporto, ...	17,474	2,339
Palermo, ...	7,983	57,115
Genoa and Leghorn, . .	635,121	399,674
Trieste, ...	165,175	3,000
Alexandria, Egypt, . .	4,000	———
Lisbon, ...	167,195	64,662
Canary Islands, . . .	3,358	5,125
Madeira, ...	———	400
Bilboa, ...	2,500	———
China and East-Indies, .	34,388	36,942
Africa, ...	25,195	12,230
Australia, ...	377,884	304,166
Otago, N. Z., . . .	10,810	5,500
Sidney, N. S. W., . .	97,880	43,012
Brazil, ...	149,676	160,152
Mexico, ...	112,986	69,481
Cuba, ...	418,034	356,436
Argentine Republic, . .	20,260	24,470
Cisalpine Republic, . .	78,552	117,626
Chili, ...	92,550	66,550
Peru, ...	169,061	256,407
British Honduras, . .	6,072	440
British Guiana, . . .	7,881	15,104
British West-Indies, . .	70,976	60,931
British N. Am. Colonies, .	28,902	16,995
Danish West-Indies, . .	8,463	31,503
Dutch West-Indies, . .	26,638	12,143
French West-Indies, . .	16,020	9,104
Hayti, ...	7,088	12,064
Central America, . . .	993	456
Venezuela, ...	28,583	15,455
New-Granada, . . .	57,490	107,837
Porto Rico, ...	20,026	59,439
Total gallons, . .	21,288,499	19,547,604

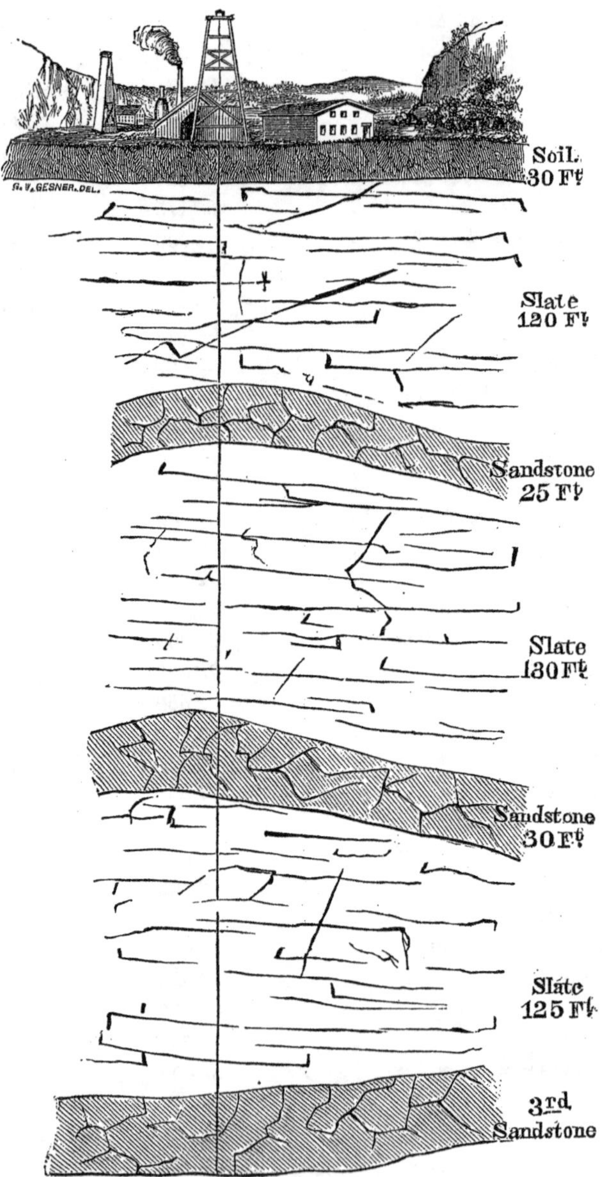

Section of Petroleum Well, Oil Creek, Pennsylvania.

The following is the quantity exported from other ports, January 1 to December 31:

		1864.	1863.
From Boston, .	Gallons	1,696,308	2,043,431
From Philadelphia, .	"	7,760,148	5,595,738
From Baltimore, . .	'	929,671	915,896
From Portland, . .	"	70,762	342,082
Total . .	"	10,457,188	8,703,117
Total export from U. S.,	"	31,755,687	28,250,721
Same time in 1862, .	"	gallons	10,887,701

The diagram opposite will serve to give some idea of the rocks pierced by the oil wells and their relative situation. The depths at which the sandstone is met vary, as also does the thickness of the slate. The third sand rock of the oil region is found to be the most fruitful in oil. In this rock the greatest flowing wells have been found, though oil frequently accompanies the boring all the way from the first sand rock. The usual depth at which oil is found is at 400 to 600 feet. Many deeper wells are in progress of boring. One is stated to have gone down 1200 feet.

MODE OF LEASING OIL LANDS.

In most cases the landowner leases for 30 years the right to raise the oil, and such right of way as may be necessary, for a royalty of from one-tenth to one-half the oil. In certain cases the landowner also receives a bonus for the lease, amounting to from $5,000 to $10,000.

The lessee binds himself to begin operations within sixty days. The lease is forfeited if the lessee fails to prosecute the work, either to success or abandonment, within a reasonable time after beginning.

Some landowners prefer to receive a certain sum for the lease, paid upon its delivery.

The value of petroleum lands upon which indications of petroleum have been found, is from $200 to $1,000 per acre in Western Virginia.

In Venango County, Pennsylvania, the heart of the oil region, the lands upon the creeks are worth $1,000 per acre. In other places not yet well known as good oil localities, the price per acre is from $100 to $300.

The prices here mentioned are those ordinarily quoted. There are instances where an acre of ground in the oil region has sold for a far greater amount than any sum named.

Previous to the discovery of petroleum, and of its value, these lands were worth in general about $20 per acre, and in some places much less.

A visit to the oil region is full of interest to most persons. There is a novelty in the operations of the oil well sinker, and in the various phenomena which attend his work, which is attractive. A ride down Oil Creek, from Titusville to Oil City, though usually over a very bad road, repays in the information gained. The derricks which in some places on the flats are almost as numerous as the trees originally were, give a peculiar appearance to the landscape. Evidences of great industry and enterprise are seen in the steam engines, vats (some of them of great capacity), barrels, boats loaded with oil floating down the creek, and many other things which serve to make up life in the oil region.

The usual mode of transportation of the oil is in barrels of 40 gallons each. Recently it has been suggested, that the oil could be forced through iron pipes from such parts as were at great distance from the railway, and a company has been formed to carry out the enterprise.

SINKING PETROLEUM WELLS.

The site of the proposed well being selected, a frame of timber called a "derrick," forty feet in height, formed of four posts inclining towards each other at the top, set upon a frame of logs eight or ten feet square or sunk into the ground, is erected over it. These derricks are prominent objects in the landscape of the oil region. (See Frontispiece and vignette.) At the top of the derrick is a pulley over which a rope is passed when the boring tools or tubing are to be hauled up. A steam-engine, usually of six or ten horse power, is placed near the derrick, and its power is applied by means of a belt from the fly-wheel of the engine to a large wheel and crank, the crank giving motion to a walking-beam, at the end of which boring tools or pump rods are attached as may be required. The lower part of the derrick is sometimes closed in with boards. The driving of the soil-pipe, Fig. 1, is the first thing done. This pipe is four inches in diameter, made in ten feet lengths, fitted at the ends, and driven by means of a heavy block of wood, as in pile driving. The lengths follow each other in succession until the rock bed is reached. This is sometimes thirty feet below the surface, and the soil pipe has to penetrate earth, sand, and loose slate, &c. The drilling tools are attached to each other

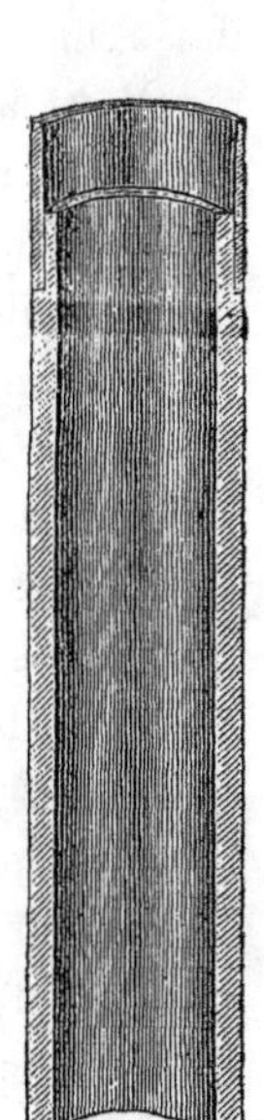

FIG. 1.

by means of a screw connection in the following order: Bit, Fig. 2, or Reamer Fig. 3; Auger Stem Fig. 4; Jars Fig. 5; Sinker Bar, similar to Auger Stem, though not so long; Rope socket Fig. 6, to which the rope is attached at one end and at the other to the Temper Screw, Fig. 9. The annexed diagram gives the position of the engine, walking beam, and connection, as they are commonly arranged, together with the relative positions of the boring tools when in use. The Temper Screw is attached to a rope which connects with the end of the walking-beam, and serves to regulate the descent of the drill, without the inconvenience of lengthening the rope at short intervals. The Sinker Bar gives weight to the upper part of the Jars, which slide together, and the Auger Stem and Bit afford weight to do the drilling. The downward stroke of the walking-beam releases the Auger Stem and Bit for an instant as the Jars slide together, and they fall the distance necessary to penetrate the rock, and are again lifted by the Jars on the upward stroke, falling again as the stroke descends.

The soil-pipe is cleared of its contents by the tools and Sand Pump, Fig. 7, which is a hollow tube with a valve at its lower end. This permits it to fill with the finely pulverized detritus made by the drill, and it is alternately lowered into and raised from the well and emptied until the well is clear for the tools again. When the soil-pipe is clear, drilling the rock is begun in the way described. The weight of an ordinary set of tools is 900 to 1000 lbs. A circular motion is given to them by means of a stout lever passed through the rope near the Temper Screw at the top of the well, and moved around gradually by one of the workmen. The Reamer is used to enlarge the hole made by the Bit.

Occasionally in drilling, the tools will enter a crevice in

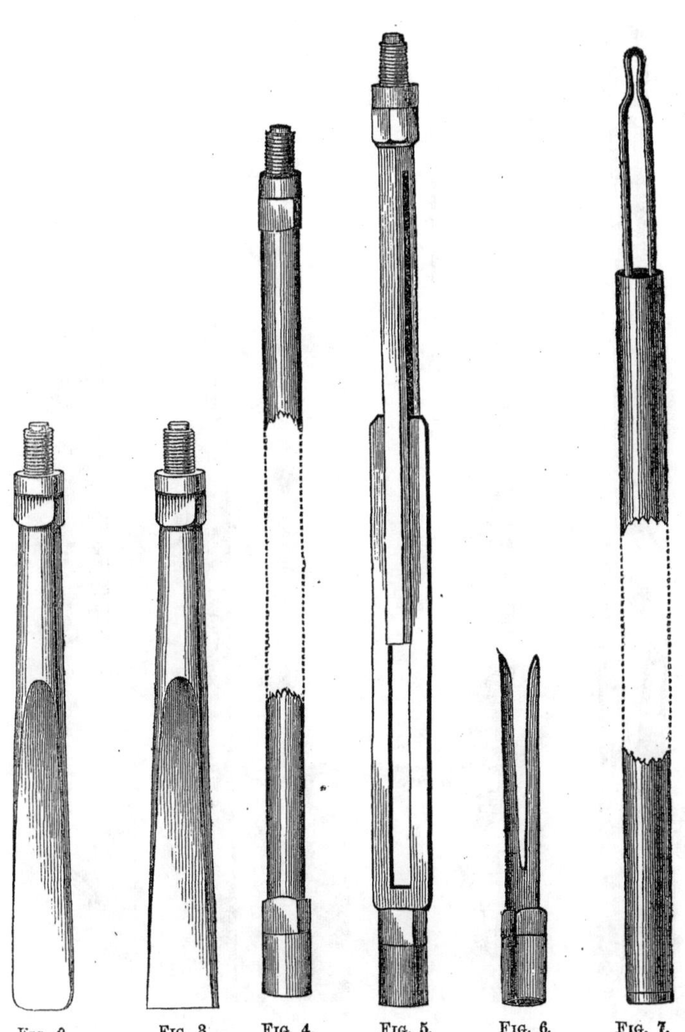

Fig. 2. Fig. 3. Fig. 4. Fig. 5. Fig. 6. Fig. 7.

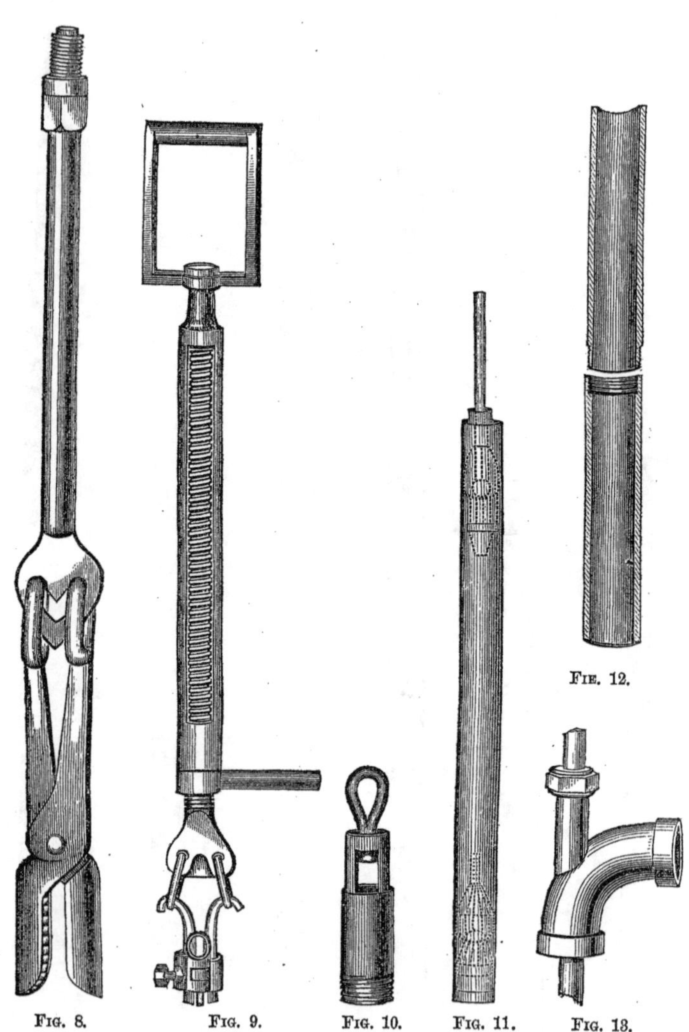

Fig. 8. Fig. 9. Fig. 10. Fig. 11. Fig. 12. Fig. 13.

Machinery used in Boring Petroleum Wells.

Boring Tools in Operation.

the rock and become wedged so tightly that they are often lost for want of means to extricate them. Numerous appliances have been invented to extricate tools that have become fast. The Lazy Tongs, Fig. 8, is one of these. It is attached by a screw-joint to the sinker bar or other suitable rod of iron, and lowered so as to catch the end of the missing tool in its jaws. It is said by the workmen that to sink one or two hundred feet is comparatively easy, but that at four hundred feet the risk of loss of tools is much increased. At that depth the jar of the falling bit and auger stem is not nearly so perceptible at the top of the well, the elastic rope taking it up.

The well having been sunk to the oil, which does not always manifest itself by flowing, it is tubed with two-inch wrought iron pipe, Fig. 12, fastened together by screw-joints, in ten or twelve feet lengths.

The first length sent down is the brass or iron cylinder which constitutes with its valves the Oil Pump, Fig. 11. This is the same size as the well tube, and has at the bottom a ball valve which is fitted into a brass plug having a screw top. The pump-rods, which are tough wooden rods fitted together by iron sleeves and screws, connect at the lower end with the upper valve of the oil pump, which has a screw socket at its lower end so that it can be lowered down to the bottom valve, screwed over its top, and pull it up when necessary. The dotted lines in the oil pump, Fig. 11, show the position of the valves, when the up stroke is at its highest pitch. When the oil pump is adjusted, the required lengths of tubing are connected one by one, and the tube is lowered to its place. In some wells, a strainer of wire cloth is placed around the lower part of the oil pump. The oil pump does not rest upon the bottom of the well so as to close its lower end. The pump-rods work through a Stuffing Box, Fig. 13, which is screwed to the top of the well tube. From its upper orifice, the oil is led by other tubes into vats, which are, in some cases, very large. In these vats the oil is settled; the water being drawn off, and the oil barrelled for market.

To prevent communication between any particular portion of the well and the pumping tube, a bag of linseed, called a "seed bag," is sent down to the required place. This bag, encircling the tube, soon swells by the water which is always present, and forms a water-tight joint in the well. For instance, when it is desirable to prevent the water beds of one of the upper rocks from flooding the oil beds of the lower strata, the seed-bag is inserted below where the water is supposed to be, and prevents it from reaching the oil pump at the bottom of the well. For drawing the tube, a Swivel, Fig. 10, is used. It screws into the tube.

The whole process of sinking petroleum wells is similar to that of the artesian wells. The peculiar requirements of the oil wells have, however, made their sinking a work of no small skill. The work of the oil-well borer is one requiring great patience and ingenuity. He works beneath the surface, where the eye cannot perceive the causes which impede the work. He has but the narrow bore of the well in which to operate, and cannot at a glance take in the whole state of the rocks through which he penetrates. His "indications" of petroleum can only be judged of by his former experience. He cannot tell to a certainty that he will "strike oil," though the sand pump may bring up traces of it from time to time. Many wells afford traces of oil, which have never yet reached any paying quantity.

When the soil is not deep, a circular excavation is made down to the rock bed, and a hollow log, or "gum," as it is called, is placed in it on one end. The base is surrounded with clay to prevent the influx of surface water. The seed bag is also used to keep the lower part of the well free of water. The cut on page 35 represents one of these comparatively shallow wells in Western Virginia. It is 150 feet in depth, and was pumping five barrels per day at the time of the author's visit in 1863.

When steam-engines are not to be had conveniently, a spring pole is used to give the proper motion to the boring tools. It is worked by two men, while another turns the lever at the top of the rope and gives a circular motion to the bit below.

There are instances in which, when the oil-well sinker is so fortunate as to tap a reservoir of petroleum, an enormous flow takes place for a time. When this is the case the pump is not required. Occasionally water is met with which flows in great quantity for a time.

Many improvements in drilling and pumping have been suggested, and a few of them are now about to be applied. An air-pump is used at some places to force a current of air through a tube carried to the bottom of the well tube, and in this way compel the oil and water to flow out without the use of pumping rods and valves. The improvements in drilling are designed to remove the detritus without the inconvenience and loss of time in using the sand pump, and to bore the well with an auger instead of a trip-hammer motion. There will be some difficulty in pulverizing the detritus so fine as to float it away on a current of water, which one invention professes to do; and also in working a long tube with auger action without torsion, which is the intent of another invention. Under favorable circumstances the wells are sunk at the rate of six and eight feet per day. There are drilling machines which can bore nine feet per hour. The difficulty will be to apply them to deep wells.

A drilling apparatus has been invented which has the bit shank hollow. The bit itself has three cutting edges formed by stout pieces of steel, set so as to radiate from a common centre. In the angle formed by these pieces, which are three inches deep, and where they join the hollow shank or stem, there are brass valves which permit the detritus to enter the stem. In this way it is proposed to combine drill and sand-pump, and save the time now taken to withdraw the tool before the pump is inserted. The above drill is, with its auger stem, worked by a wire rope, which is passed through an upright shaft at the top of the well. This shaft is furnished with set screws to catch the

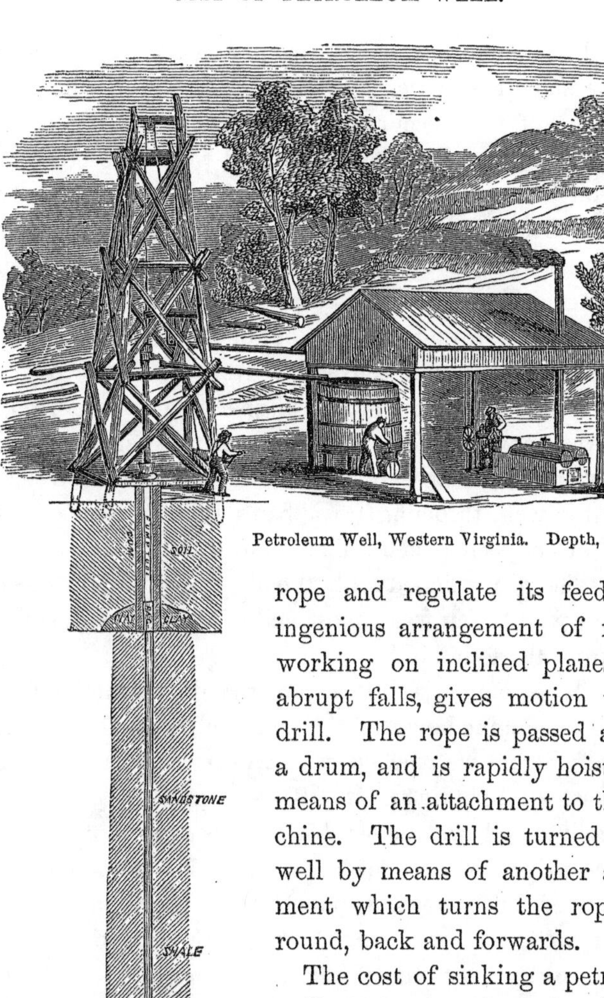

Petroleum Well, Western Virginia. Depth, 150 feet.

rope and regulate its feed. An ingenious arrangement of rollers, working on inclined planes with abrupt falls, gives motion to the drill. The rope is passed around a drum, and is rapidly hoisted by means of an attachment to the machine. The drill is turned in the well by means of another attachment which turns the rope half round, back and forwards.

The cost of sinking a petroleum well 600 feet is estimated at $7,000, but an allowance must be made for contingencies, loss of tools, etc.

Before the nature of petroleum and its inflammability had become known, several serious accidents occurred from the gas, which accompanies the oil in vast quantity and great force, taking fire. A person who witnessed one of these occurrences, thus described it:

"We had gone down 300 feet and were expecting to strike oil at any moment. We went up to the shanty where we boarded to supper, and on our way back to the well, which was just below in the hollow, we saw the men hurrahing, and presently a jet of gas, water, and oil, rushed up, fairly lifting the tools out of the well. It roared and hissed like letting off steam from a boiler. The stream seemed to me to mount higher than the derrick, which was forty feet high. The folks in the neighborhood ran down with their shovels and dug a circular trench around the well, throwing up a bank to catch the oil, as we had not expected such a flood, and had no large tanks ready. The gas mingled with the air, and for a distance about the well the air was almost yellow with gas and spray of oil from the fountain.

"Mr. R—— and myself looked on awhile and then started to go to the engine-house of the next well to have the fires put out. Before we reached it, however, the gas took fire like a flash of lightning. Mr. R——, who was passing a small tank of oil, was covered with it as it took fire also, and I lost sight of him for a moment. My hair and face were burned, but I was not much hurt. The sight of the burning well was terrible. A great fountain of fire, it wavered to and fro as the wind took it, and threw off blazing jets of oil. The poor people who were dipping the oil up in the little pool around the well, wilted down like leaves when the forest is on fire. Some tried to crawl away, but the liquid flame ran along the ground and caught them. Several hundred barrels of oil from a neighboring well caught fire. Vast clouds of smoke rose from the burning well and floated off over the hills, and when night set in the clouds and hills were red with the light of the conflagration. Mr. R—— died very soon after. There were a great many lives lost."

Petroleum should always be handled carefully. Though it will not explode by being fired, yet the consequences resulting from the sudden inflaming of a large quantity of a fluid so highly combustible, are in many cases more destructive than those of an explosion of gunpowder.

Recently, in Philadelphia, a large quantity of petroleum, stored in a street near some dwelling-houses, was fired at night. The store yard was so situated that the oil found an easy descent by the street, and flowed down it in a river of flame, running into the cellars and setting the houses on fire with the rapidity of a train of gunpowder.

In warm weather there is a considerable volatilization of the lighter portions of petroleum. These will fire instantly from a match. Barrels which have contained petroleum or coal oil will sometimes become filled with just such a mixture of gas and air as to be very explosive. A negro, at a small town in Georgia, sitting upon an empty kerosene barrel, lit a match at the bung. The barrel exploded, blowing out the heads, but the negro was fortunately not much hurt. Accidents from the gases of petroleum and coal oil occur also in stills which have not been ventilated properly before being approached with a light.

VARIETIES AND PRODUCTS OF PETROLEUM.

The oil wells of Pennsylvania yield generally greenish oils of rather unpleasant odors. Their specific gravity ranges from ·820 to ·782, or from proof 40° to proof 48° Baumé. The oil by distillation yields from 75 to 85 per cent. of a lamp oil which should not vaporize and inflame under a temperature of from 110° to 116° Fah. The refined oil is usually sold subject to the above test.

The heavy oils, or residuum, left in the still are subjected to the paraffin treatment, and sold as lubricators. In some cases the paraffin is not separated, but the whole residuum is mixed with various matters and sold as cart grease. The naptha, or benzole, as it is improperly called, is used as a substitute for spirits of turpentine, or is mixed with turpentine for painting purposes. The naptha is obtained in varying proportions from the petroleum, but is usually 10 to 20 per cent. of the crude oil.

A very heavy lubricating oil is obtained from several wells in Ohio and Pennsylvania, of specific gravities ·880 to ·860, or from proof 28° to 32° Baumé. Some of these heavy oils will remain fluid at very low temperature. The "Mecca oil" is of this class. Its specific gravity is from ·890 to ·910, or from proofs 23° to 26° Baumé.

An oil is obtained from the wells of the Tarentum Oil, Salt, and Coal Co., of the specific gravity of ·795, or proof 45° Baumé. It is of a dark amber color, and will burn for a time in lamps without being previously refined. It yields 5 per cent. of naptha by distillation, and 90 per cent. of lamp oil. It is used by woollen manufacturers in place of lard oil.

A very heavy oil is obtained near Crestline, Ohio. It resembles the Mecca oil, and is an excellent lubricator. The wells at Franklin, Pennsylvania, produce heavy oils. An oil, which is quite equal in color to the best refined paraffin oil, is found on Duck Creek, Ohio.

The Canada petroleum is of specific gravity from ·880 to ·860, or from proof 28° to 32° Baumé. It is a dark colored and offensive oil. Its odor can be removed as shown in another place. It yields more lamp oil than the Pennsylvania, as it will burn in a lamp at proof 36° Baumé, or specific gravity ·838.

The California petroleum varies in density. Mr. J. H. White, of San Francisco, gives the yield of the petroleum tested by him as below, the crude oil being at 20° Baumé, or specific gravity ·927:

38 per cent.	Illuminating oil, . . .	41° Baumé.
48 "	Lubricating oil, . . .	21° "
10 Pitch,		
4 Water.		
100		

Mr. Gilbert, who has had some experience in California petroleum, states that the crude oil loses 10 to 15 per cent. of its volume in the process of rectification, and classifies the products as follows:

Light oil (Naphtha) . .	5	per cent. at	65°
Burning oil "	50	"	30° to 32°
Light machine oil, . .	20	"	25°
Heavy oil and paraffin, .	25	"	18°

The above is taken from Prof. Silliman's report on the California Petroleum region. The proofs mentioned are, no doubt, those of Baumé's hydrometer. The density of the illuminating oil seems to be stated rather low, but it may be that the lamp oil from California petroleum can be burned at a lower proof than even that of Canada, which burns at 36° Baumé.

There has been some difference of opinion regarding the yield of California petroleum in lamp oil. Both of the oils examined by Mr. White and Mr. Gilbert, are certainly valuable. In a region where petroleum can be had in a variety of conditions, samples can be tested which would bear no comparison in product to each other.

Hardened petroleum from Enniskillen, Canada West, was tested by the author and found to yield 50 per cent. of products by distillation. A sample not quite so indurated gave 65 per cent.

Messrs. Parsons and Conway report that crude oil from the Buenaventure district, California, produces 50 per cent. of lamp oil, and 28 per cent. of lubricating oil.

CANADA PETROLEUM.

The petroleum wells of Canada are situated principally in the county of Lambton, Canada West. The oil springs of Enniskillen, and of the banks of the Thames, were long ago known to the Indians and early settlers. On Black Creek, petroleum springs had in course of time covered an area of two acres with semi-solid bitumen, the result of its loss of the lighter portions by evaporation.

At Enniskillen the rock bed is covered with from forty to sixty feet of clay and a thin bed of gravel.

The "surface wells," so called, are sunk through the clay to the rock, in the same way that an ordinary shaft is sent down, care being taken to curb the shaft with plank or timber, to prevent the caving in of the soft and water-laden sides. Upon reaching the gravel bed, the petroleum is generally met with in considerable quantity, though it frequently appears in the shaft as it descends. The clay seems to be filled with veins of water and petroleum, and from these the surface wells are supplied. The "rock wells," as they are termed, are those deeper borings which resemble those of Pennsylvania. They are sometimes sunk when the surface well becomes exhausted, by continuing the wells by the drills, or are commenced, as in Pennsylvania, by driving to the rock-bed.

At a distance of two hundred feet from the surface the petroleum is found; sometimes in great abundance. Usually petroleum and water are produced by the well, but recent wells have been found to send up petroleum only.

The townships of Mosa and Oxford, on the Thames, and a locality on the Big Otter Creek, in Dereham, near Tilsonburg, have been found to contain the oil. Wells have been sunk at Gaspè, Canada East, near Douglastown, where petroleum springs are found. So far, they have not been found to yield very largely. Indications of petroleum, however, exist over a large portion of Gaspè, and it is very probable that a large supply may yet be had from this quarter.

PETROLEUM OF TRINIDAD.

The celebrated Pitch Lake of the Island of Trinidad is upwards of three miles in circumference, and forms the head of La Brae harbor. At the time of the author's visit to the place, the bitumen, of the consistency of thin mortar, was flowing out from the side of a hill, and making its way outwards over more compact layers towards the sea. As the semi-solid and sulphureous mineral advances, and is exposed to the atmosphere, it becomes more solid; but ever continues to advance and encroach upon the water of the harbor. The surface of the bitumen is occupied by small ponds of water—clear and transparent, in which there are several kinds of beautiful fishes. The sea, near the shore, sends up considerable quantities of naphtha from submarine springs, and the water is often covered with oil, which reflects the colors of the rainbow. In the cliffs, along the shores, there are strata of lignite, in which it has been supposed by some the bitumen and naphtha had their origin.

PETROLEUM OF CUBA, WEST INDIA ISLANDS, AND SOUTH AMERICA.

In Venezuela, at the Punta d'Araya, at Cape Cirial, and near Cape de la Brea, Von Humboldt observed "streams of naphtha issuing from mica slate, and covering the sea for some distance."* At the Lake, and near the city of Maracaybo, numerous streams of petroleum are found, together with compact bitumen.

Petroleum springs have also been discovered in Brazil. Most of the West India Islands contain petroleum and bitumen.

R. C. Taylor, in his "Statistics of Coal," 1855, states that from the serpentine rock at Guanabacoa, near Havana, petroleum springs are observed issuing. There are all the varieties of bitumen to be found here, from the thin oil to the compact pitch. It is a matter of history, that Havana was originally named "Carine" by the early visitors and settlers; "for there we careened our ships, and pitched them with the natural tar which we found lying in abundance upon the shores of this beautiful bay."†

"There are also springs of petroleum between Holguin and Mayari, in the eastern part of Cuba, and also in the direction of Santiago de Cuba."‡ There is a petroleum spring

* Travels and Researches of Alexander Von Humboldt, 1799.

† Early History of Cuba.

‡ Essai Politique sur l'Isle de Cuba.

in St. Andrew's parish, Barbadoes. The product of this spring has been sold under the name of "green tar," and "Barbadoes tar."

PETROLEUM OF BURMAH, JAVA, ETC.

The celebrated Rangoon wells at Yananhoung, on the Irawaddy, produce yearly a large supply of petroleum, which is regularly imported into England. It contains from ten to eleven per cent. of paraffin.

Mineral oils are also found in many islands of the Indian Archipelago. Java and Sumatra produce them. An almost colorless oil rises from the earth at Baku, on the borders of the Caspian Sea. It can be burned in lamps in its natural state. Petroleum is found at Schnde, near Hildesheim, in Hanover, and in Gallicia; at Amiano, in Italy; in Bavaria, Sicily, Switzerland, France, and in other parts of Europe.

It would seem from the numerous places on the globe where petroleum exists, that it must for a long time do away with the necessity for coal distillation. But though petroleum has been discovered at the places mentioned, it has been produced as an article of commerce at but a very few of them, and it is not probable that from many of them any large supply will ever be obtained. There will always be places where coal and coal shales will be the most profitable to the manufacturer. In Great Britain, for instance, the distillation of shales is steadily increasing, though its commerce with the United States is very large, and the shipments of petroleum from the latter country are extensive.

CHAPTER III.

Coal—Bituminous Clays and Shales—Bitumen—Table of Volatile Matter, Coke and Crude Oil from Coals, etc.

THE varieties of coal have heretofore been classed under the heads of

Anthracite, or Hard Coal,
Caking Coal,
Cherry Coal,
Splint Coal, and
Cannel Coal.

These five varieties have the following composition:—

	Richardson.		Thompson.		
	Anthracite.	Caking Coal.	Cherry Coal.	Splint Coal.	Cannel Coal.
Carbon	92·56	87·952	83·025	82·924	76·25
Hydrogen	3·33	5·239	5·250	5·491	5·50
Nitrogen	"	"	"	"	1·61
Oxygen	2·53	3·806	8·566	8·847	13·83
Ashes	1·58	1·393	1·549	1·128	2·81

Other varieties of combustibles have been arranged by Berthier in the following manner:—

Composition in 100 parts.	Peat or Turf.	Lignite, or Brown Coal.	Bituminous Coal.	Anthracite, Pennsylvania.	Plumbago, or Graphite.
Carbon	38	54	73	94	95
Hydrogen	"	05	05	2·55	"
Oxygen	"	26	20	2·56	"
Ashes	17.4	14	02	"	"
Volatile Matter	28	"	"	"	"
Iron	"	"	"	"	5

The names given to combustible substances have been applied with reference only to their common characters and uses. Frequently coals bear the names of the places where they are mined. Few of their appellations have any reference whatever to their chemical composition, and therefore in seeking for *oil coals* (if the term may be used) the manufacturer must be directed by an actual test of the material itself.

In the same coal field, the same series of strata, and in the same stratum, there are important differences of composition. It is as providential as wonderful that the carbonaceous material of the same deposit is adapted to different uses.

The varieties of coal may have been produced from the different kinds of plants from which the coal has been derived, and the peculiar conditions of the districts where those plants flourished before their downfall and inhumation, or submersion. The changes that have taken place in the original plants during their passage from woody fibre into coal, are ascribed to the evolution of a part of their carbon, hydrogen, and oxygen, as there is less of those elements in the coal than in wood. This will be observed by viewing the following table:—

	Carbon.	Hydrogen.	Oxygen and Nitrogen.
Composition of the Anthracites of the *Transition Rocks* .	90·	2·50	3·69
" " Bituminous Coal of the *Secondary Rocks* .	86·	4·86	7·11
" " Lignites of the *Tertiary Rocks* . . .	60 36	5·00	25 62
Wood (recent)	49·60	5 80	42·56

It will also be observed that the older the formation the greater the amount of carbon contained in its coal, the amount of hydrogen being diminished. This fact may be ascribed, chiefly, or in part, to the greater degree of heat and pressure to which the lower and older coal strata have been, and still are subjected.

The gases of deep coal mines are very similar to those of gas manufactories, and such as are produced by a high temperature. The deeper the mine the greater is the discharge of carburetted hydrogen. It is to the internal heat of the earth, and other chemical agencies combined with causes of less force, that we must chiefly ascribe the transmutation of wood into coal. The similarity of the distilled products of wood and coals, and of charcoal and coke, should not be overlooked in seeking for proofs of the vegetable origin of coal. In mines of lignite and cannel coal carbonic acid or *choke damp* is almost the only gas present. In lower coal mines, or those that have been longer under the influence of heating and other chemical agents, carburetted hydrogen, or *fire damp*, predominates.

Liebig has shown the great loss of oxygen and increase

of hydrogen and carbon in lignite and brown coal, during their transition from a vegetable to a fossil state; still there is much that remains unexplained regarding other kinds of coal.

BOGHEAD COAL, OR BITUMINOUS CLAY.

This peculiar mineral occurs at Torbane Hill, in the carboniferous limestone of the Frith of Forth, Scotland. It is the material from which Mr. Young obtains paraffin oil and paraffin, and his manufactory is in the immediate vicinity of the mines. It has been extensively shipped to the United States, and employed in the manufacture of kerosene at New York and Boston. During the year 1859 the North American Kerosene Gas Light Company imported upwards of 20,000 tons of this material for the supply of their works at Newtown Creek, Long Island, and at an average cost of eighteen dollars per ton. In consequence of the discovery of numerous strata of cannel coals in the Western States of this country, and of cheaper substances for the production of oils, the importation of the Torbane Hill mineral will doubtless be discontinued.

Although this mineral possesses few of the characteristics of a true coal, the term coal has been applied to it for commercial convenience. It has been the source of long-continued and expensive lawsuits. The point in dispute affected the ownership, whether it was coal, or not coal. Numbers of the most scientific men in Europe were arraigned before courts and juries to decide whether the so-called Boghead coal is coal, or bituminous clay. There was a decided preponderance against the term "*coal*," and in favor of "bituminous clay." Finally the contending parties compromised, and the term coal was permitted to be applied, although the bitumen of the Great Pitch Lake of Trinidad has an equal right to that appellation.*

Boghead coal is among the most valuable minerals for the manufacture of oils. It has an uneven fracture, is of an earthy color, and burns with a long smoky flame. It yields 13,000 cubic feet of gas, of specific gravity 0·775 per ton. As it contains only traces of nitrogen, the quantity of ammonia given off is small. The following is the medium result of four trials in testing its qualities :—

Volatile matter	70·10
Carbon in coke	10·30
Ash	19·60
	100·00

The ton of coal run in common retorts gives 120 gallons of crude oil, of which 65 gallons may be made into lamp oil, 7 gallons of paraffin oil, and 12 lbs. of pure paraffin. The coke is worthless, and the ash consists chiefly of silica and alumina. At a price of 11 dollars per ton for the coals, the cost of the oil is estimated at 63 cents per gallon.†

ALBERT COAL.

This bituminous mineral occurs at Hillsboro', Albert County, in the province of New Brunswick, and within four miles of the Peticodiac River. It is an injected vein, situated almost vertically in the earth, and from one to sixteen feet in thickness. It is associated with rocks highly charged with bitumen, and has neither roof, floor, under-

* *London Journal of Gas Lighting,* iii. 521. Young *vs.* White, and others.

† Report of the Committee North American Kerosene Gas Light Company. New York. 1860.

clay, nor stratum of *stigmaria*, nor other accompaniments which distinguish coal deposits from all others.*

The Albert coal, so called, is extremely brilliant, breaks with a conchoidal fracture, does not soil the fingers, and is strongly electric. It melts, and drops in the flame of a candle, and dissolves in naphtha and other solvents, forming a varnish. It has all the essential properties of asphaltum, while it is void of those which constitute true coal. Like the mineral of Torbane Hill, it has been the subject of disputes and lawsuits, the total cost of which has exceeded thirty thousand dollars. If the substance were coal, the coal was the property of one party; if asphaltum, the asphaltum belonged to another. Coal had been reserved by the Crown of Great Britain; but asphaltum was not mentioned in the grants of the land. In April, 1852, an intelligent jury, who analysed the mineral at Halifax, decided that it was asphaltum, and not coal. Another trial was held in the county where the so-called Albert coal is mined in July of the same year. It lasted eleven days. Chemists were summoned from every quarter, and under the most conflicting testimony, and with a jury of farmers, the advocates for coal obtained a verdict, and the asphaltum has since been called Albert coal. The composition of the Albert coal is as follows:—

Carbon	86·207	85·400
Hydrogen . . .	8·962	9·200
Nitrogen . . .	2·930	3·060
Sulphur . . .	traces	a trace
Oxygen . . .	1·971	2·220
Ash	0·100	0·120
	100·000	100·000
	C. M. Wetherell.	Gesner.

* See Taylor on Coal, 2d edition, p. 516.

The average yield of crude oils by four trials in large retorts was 110 gallons per ton, and

Volatile matters	61·050
Coke	30·650
Hygroscopic moisture	0·860
Ash	7·440
	100·000

Of the crude oil 70 per cent. may be made into lamp oil, 10 per cent. is heavy oil and paraffin. The coke is exceedingly brilliant and cellular; it burns rapidly, and gives a strong heat.

BRECKENRIDGE COAL.

The Alleghany, or Apalachian coal field of the United States, has been estimated to embrace 63,000 square miles. Interstratified with the common bituminous coals, in this vast region there are very numerous strata of cannel coals, adapted to the manufacture of oils. In the numerous surveys and valuable reports made on the coal districts, cannel coals are seldom described as a distinct variety.

A peculiarity of the great Western coal field is, that the coal does not appear to be separated into basins, or lake-like depressions in the earth, as it is in Europe, and in the anthracite coal districts. The bituminous coal is found in the tops of hills, and even in the Alleghany Mountains, in beds nearly horizontal, and it displays the same peculiarity as it stretches away towards the Gulf of Mexico, the Canadian Lakes, and the Rocky Mountains.

Among the cannels that have been discovered Breckenridge coal holds an important place. This coal occurs in Breckenridge County, Kentucky. It is a rich variety of cannel, three feet in thickness, and has already supplied a large amount of oil and paraffin. The lamp oil, when properly purified, is of good quality. At a red heat this coal yields—

Volatile matters	61·300
Fixed carbon	30·000
Ash	8·055
Hygroscopic moisture	·645
Sulphur	a trace
	100·000

By the ordinary methods of working, this coal yields 130 gallons of crude oil per ton, of which 58 per cent. was manufactured into lamp oil, and 12 gallons into paraffin oil and paraffin. The quality of the coal is variable, and the products are very much influenced by the degree of heat applied to the retorts in the distillation.

CANDLE TAR.

The tar and pitch resulting from the manufacture of stearine have been employed for the production of oils. Large supplies have been imported from England into the United States, and sold under the names of "grease" and candle tar. The ordinary yield of crude oil from this material is 200 gallons per ton, of which 70 per cent. may be made into lamp oil, and 10 per cent. into lubricating oil. The oils are excellent in quality; but heretofore the first distillation of the tar has been attended with inconvenience, as it "foams up" in the retorts, and the coke adheres very firmly to their sides. The price has varied from 25 to 40 dollars per ton.

This article was distilled in New York by Henry Gesner,

in 1857. The tar foams in the beginning of the distillation, and is very troublesome. The foaming arises from water intimately combined with the tar, and causes the charge to "boil over." In one instance, the expansive force was so great as to lift the dome of the still, which was let into a groove around the sides, and not bolted, but cemented. The consequence was that the charge in the still spewed out and took fire, and the factory was destroyed. There is a small portion of stearic acid distilled over, after the water has been removed. The distillate runs from 65° Baumé to 30° Baumé. There is a hard coke or pitch left in the still.

The residuum of palm oil and lard distillations behaves in much the same way. The illuminating oil is excellent.

SOUTH BOGHEAD COAL.

Near Poole (England) there occurs a peculiar kind of shale, which has been sold as "South Boghead Coal." It abounds in the remains of fishes and *crustacea*. It gives out 42 per cent. of volatile matter, and therefore has offered an object of trial to oil-makers; but the oils made from this rock contain a greater number of the equivalents of carbon than those derived from coals, or bitumens, and with the ordinary density they smoke in the common lamp. It seems quite evident that the elements of the oil—carbon, hydrogen, oxygen, and nitrogen, now composing the shale, in part, have been derived from fishes and other marine animals, and not from the vegetable kingdom, as in the case of coal.

BROWN COAL.

Singular beds of brown coal have been discovered on the Ouachita River, Arkansas, and at other places in that quarter. They contain the remains of *sphagneous* plants and woody fibre. It appears that peat bogs have been overflown, or otherwise saturated with petroleum, and hardened by time and oxidation. The oils distilled from this material abound in paraffin. It has the following composition in 100 parts:—

Volatile matters condensed into oils, and gas uncondensed	60·10
Fixed carbon	32·85
Ash	7·05
	100·00

Crude oil at the rate of 68 gallons per ton was obtained from it. It is semi-solid when the thermometer is at 80° Fah., and, besides lamp and lubricating oils, it produces 143 lbs. of paraffin per ton.

Nova Scotia and New Brunswick produce a variety of shales, which at one time could be profitably used in the manufacture of oil. The "Pictou shale" of Nova Scotia, and the "Asphalte Rock" of Dorchester, New Brunswick, yield from twenty to thirty gallons refined oil to the ton.

A sample of cannel coal from Hunter's River, Australia, gave a yield of sixty gallons crude, and forty gallons refined oil to the ton. Australia evidently abounds in bituminous substances. Mr. H. H. Hall, of Sidney, N. S. W., handed the author several fine samples of cannel coal from the vicinity of Sidney, together with a small specimen of shale, which, in appearance and burning properties, resembled the Boghead coal.

BITUMEN AND BITUMINOUS SANDS AND CLAYS.

At the various localities mentioned in the preceding chapter as petroleum deposits, bitumen in different stages of solidity is generally found.

The bitumen of Trinidad was the article from which the author first obtained "Kerosene," which differs in some degree from "coal oil." The bitumen is of a grey color, somewhat brittle, but still yielding to the heat of the sun. A cargo of broken masses will consolidate in a ship's hold in such a manner as to require mining before it can be discharged. The following is the result of several trials made with reference to its application for the manufacture of oils:

Specific gravity . . .	0.882	
Crude oil	70	gallons per ton.
Refined oil	42	" " "
Lubricating oil	11	" " "

This bitumen varies in quality, owing to the sand and debris over which it flows. Taken from the lake itself, it would probably yield twice the above quantity of oil.

A vein of bitumen has been discovered near Cairo, thirty miles east of Parkersburg, West Virginia. It is represented as a perpendicular mass, jutting out from the side of a hill two hundred and ninety feet. The strata of the hill are nearly horizontal, and they are cut at right angles by the continuous vein of the bituminous mineral, which is four feet eight inches in thickness. The position of the vein has been ascertained by the proprietors, who have sunk a shaft upon the line of the outcrop. A sensible description represents that it appears the hill has been split, a perpendicular chasm opened, and afterwards filled with asphaltum in a liquid state, and which has since hardened into a compact material. Coal never occurs in this manner; but is always interstratified with its associate sandstones, shales, and fire clays. In all its geological relations and character, the Cairo deposit is like the asphaltum of Albert County, New Brunswick.

Some of the Cuba bitumens yield one hundred and twenty gallons of crude oil to the ton. The finest varieties are used in making varnish. On the borders of the River Acarahy, forty miles south of Bahia, there are extensive beds of lime and clay saturated with bitumen, and capable of yielding oil in large quantities. The oils resemble those obtained from Trinidad or Cuba bitumen.

The annexed table will afford some guide to the manufacturer regarding the proportion that crude oil bears to the volatile matters of the material, and also regarding the localities of coals, shales, bitumen, etc. The refined products of crude coal oil depend so much upon their treatment that it is difficult to express in figures their actual amount.

Breckenridge coal, as has been shown, gives—

130 galls. crude oil per ton.

From which we obtain	80	gallons	illuminating oil,
and . .	12	‘	paraffin oil,
Making	92	"	in all of marketable oils.

Boghead coal oil yields 120 galls. crude oil per ton.

From which we obtain	65	gallons	illuminating oil,
	7	"	paraffin oil,
and. . . .	12	lbs.	paraffin.
Equal to about . .	84	gallons	of marketable oils.

Yet by experimental distillation Boghead coal yields more volatile matter, and leaves less coke than the Breckenridge.

Locality.	Volatile Matters.	Coke.	Yield of Crude Oil per ton.
England.			
Derbyshire	48.36	53	82 gallons.
Wigan Cannel	44	56	74
Liverpool "	39	61	50
Poole (Shale)	42	58	50
Newcastle	35	65	48
Scotland.			
Boghead	70·10	29·90	120
Scotch Cannel	38	62	40
Lesmahago	51	49	96
Provincial.			
Albert Coal, N. Brunswick .	61·050	30·65	110
Asphalte Rock, " .	43	57	64
Pictou Shale, Nova Scotia .	27	73	47
American.			
Breckenridge	61·30	38·55	130
Erie Railroad	35	65	47
Newburg	38	52	72
Falling Rock	50	50	80
Pittsburg	36	64	49
Kanawha	46	54	71
Elk River	41	59	60
Cannelton	34	66	86
Coshocton, Ohio	45	64	74
Darlington, Pa.	42	58	56
Ouachita River, Arkansas .	60	40	64
Bitumen, etc., United States.			
Ritchie County, Virginia . .			170 gals. per ton.
Pennsylvania	—	—	—
Petroleum Springs, Alabama, Georgia, Tennessee, Kentucky, Virginia, Maryland, Ohio, Penns'lvania, Canada			From 75 to 85 per cent. of Lamp Oil.
Cuba.			
Bitumen	71	29	120
Trinidad.			
Bitumen	38	52	70
Canada etc.			
Bitumen	70	30	118
Illinois "Gas Stone" . . .	26	Limest'e.	18
California	70	30	116
Brazil	78	22	
Peat	71	25	6 to 8

Actual test by retorting and distilling must be made before judging of the value of any particular coal for oils.

Persons unaccustomed to handling coals may judge roughly of their value for oil by noticing their weight, lustre, etc.

As a rule, a dull fracture, great comparative lightness, and easy inflammability in the flame of a candle, are favorable signs of an oil coal.

Sometimes shales of very inferior appearance are rich in oil. Commercially, the value of any coal for oil will depend upon its situation. There are places where the distillation of shales can be profitably made by the heat afforded by ordinary bituminous coals; shales never yielding coke of any value.

A coal which will yield sixty gallons of crude oil, or forty gallons of refined oil for lamps, may be regarded as an excellent article when it will afford coke enough to supply the heat for its own distillation.

Peat has been distilled for oils in Ireland, and in Kildare the extensive works of the Irish Peat Company have been in operation for that purpose for some time. One ton of peat yields on an average:

Liquids (not oily),	65 Gallons.
Tar	6 "
From which are produced:—	
Lamp oil,	2 Gallons.
Lubricating oil,	1 Gallon.
Paraffin,	3 lbs.
Ammonia,	3 lbs.
Acetic acid,	5½ lbs.
Naphtha,	8 lbs.
And 25 per cent. of charcoal.	

Thus far, however, peat has not been a successful competitor of coal, bitumen, or other more compact carbonaceous materials.

CHAPTER IV.

Nature of the products distilled from Bituminous Substances.—Modes of obtaining Oils.—Retorts.—D-shaped Retorts.—Revolving Retorts.—Vertical Retorts.—Clay Retorts.—Brick Ovens.—Coke Ovens.—Stills.—Condensers.—Agitators.—Super-heaters.

To obtain oils from coals and other dry bituminous materials, it is necessary in the first place that they shall be submitted to dry or decomposing distillation, by which oils are formed, the coke or fixed carbon remaining with the ash in the distilling vessel. The economy and perfection of this operation depend upon the kind of retort used, the degree of heat applied, and the efficiency of the condensing part of the apparatus. If a given quantity of coal be distilled in a retort or close vessel at a heat of 1,200° or thereabouts, in the manner that coal gas is made, a large quantity of gas will be formed. The oily products will be small in quantity, and consist chiefly of benzole, naphtha, naphthalin, carbolic acid, piccamar, pittical, copnomor, and other hydrocarbons, which, so far as the oil manufacturer's objects are concerned, may be called impurities. They are not impurities; but in the progress of chemical science they may hereafter become valuable. Again, the crude oils obtained by such a heat contain more carbon than those produced by a lower heat, much of the hydrogen being driven off from the coal in carburetted gases. But if the heat to which the coals are exposed does not exceed 750° or 800° Fah., a different class of results follows. Instead of true benzole, eupion* will be formed, the naphthalin will be replaced by paraffin, the carbolic acid, piccamar, copnomor, etc., will be less in quantity, and there will be a great increase of the oils employed in lamps and for oiling machinery.

To obtain these results, not a little will depend upon the form of the retort, and the mode by which the oily vapors generated in it are condensed. The retort which will permit the charge of coal to be equally heated throughout, is best; for if the heat be strong on one part of the charge, and weak on another part, the former will produce permanent gas and impure oils, while the latter has, perhaps, a temperature too low to produce oils at all. It is on this account that revolving retorts, which keep the charge in constant motion, have been introduced.

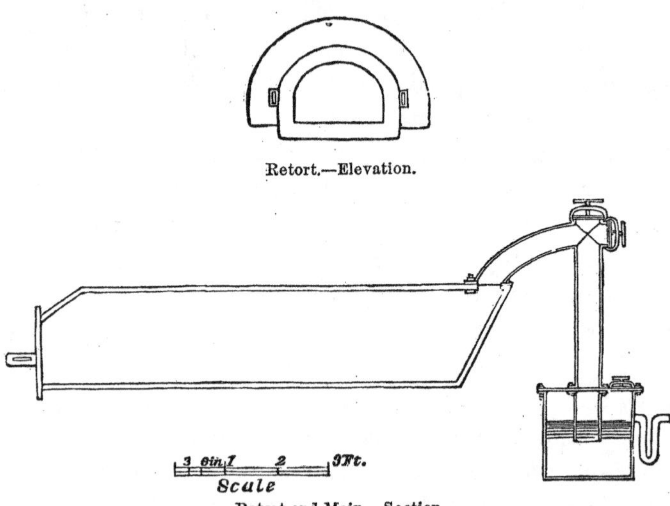

Retort.—Elevation.

Retort and Main.—Section.

Full descriptions of all the retorts and ovens which have been experimented with, tried, patented, used, and in use,

* The composition of benzole is C_{12}. H_6. That of eupion is C_5. H_6.

for distilling oils from coals, would occupy too much space. For such descriptions it is necessary to refer to the American, English, and French records of patent inventions. Great as their number is, it is still increasing. Many persons besides chemists, who are concerned in this kind of manufacture, and tyros in the art, have a fancy for some novelty in which neither philosophy nor chemistry can discover any merit, and vast sums of money have been wasted in seeking the talisman that would convert everything into oil.

Horizontal D-shaped retorts, of large size, two or three being heated over one furnace, have proved satisfactory;

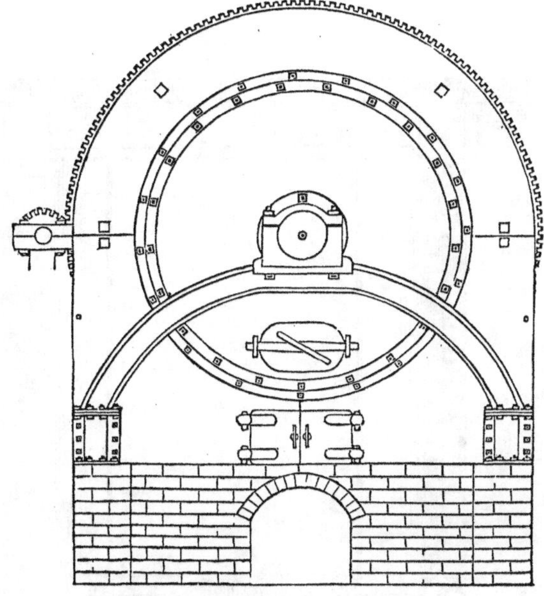

Revolving Retort.—Front Elevation.

and in some instances they have taken the places of the revolving cylinders. They may be made of iron or clay.

They are simple in construction, and readily charged and discharged. They may be made from thirty to forty-five inches in width, and from eight to ten feet in length. The latter size will distil three charges of cannel coal, of 450lbs. each, in twenty-four hours, at a heat not exceeding 780° Fah. Forty of these retorts and more may discharge into one main, from which the gas is conveyed to a gasometer, to be afterwards used for fuel or for lighting. It is necessary that the discharge-pipes leading from these retorts to the main should not be less than eight inches in diameter,

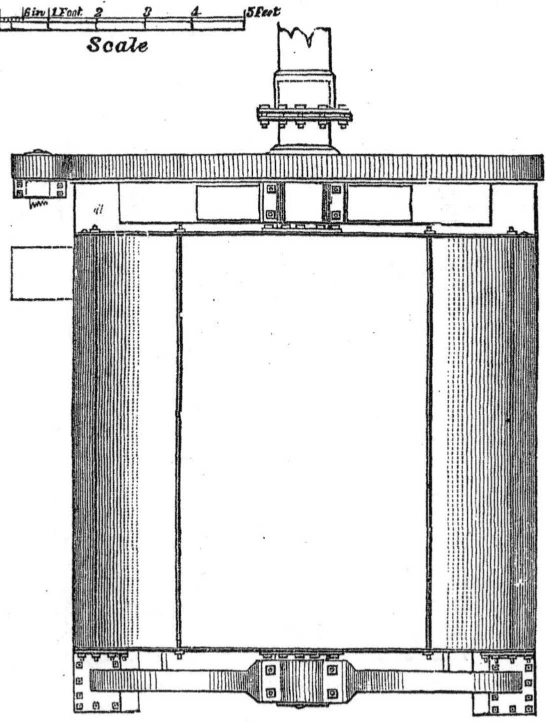

Revolving Retort.—Plan.

to prevent pressure and insure safety; and they should be

inserted into the end of the retort opposite the head and furnace, and upon a level with the upper part of the charge. The main itself should be three feet in diameter.

REVOLVING RETORTS.

Revolving retorts were employed by Gingembre, of France,* and by others, many years ago, in the manufacture of coal gas; but from their cost and liability to get out of order, they were discarded. Since January, 1858, several patents have been granted in the United States for this kind of retort, as being adapted to the manufacture of oils from coals, shales, etc. In some establishments they are now in use; in others they have been replaced by large D-shaped stationary retorts.

They are iron or clay cylinders, frequently six feet in diameter and eight feet long, sustained upon an axle at each end, the vapors passing through the axle opposite the furnace, or head, where they are charged through a man-hole in the usual manner. They are kept in motion by machinery propelled by steam, making two or more revolutions per minute. The advantages of the revolving retort are, that the charge being constantly agitated by the motion of the cylinder, every part of the material is from time to time brought in contact with a heated surface, so that it is exhausted in much less time than it could be in a stationary retort; thus, also, there is a saving of fuel. A retort of the above dimensions will run six charges of one ton each, in twenty-four hours, of ordinary cannel coals. The objections urged against them, by those who have given them a trial, are their cost and liability to get out of order. They also grind the coal to a powder, which, by

* *Brevets d'Invention*, vol. ix., p. 235.

being carried along with the oily vapors, is apt to fill up the condensing worm, and its admixture with the oil increases the cost of purification. But the rapidity by which they distil the coal, and the saving of fuel, are certain results; and the ingenuity of numerous inventors may hereafter relieve them from the above drawbacks.

The revolving retort cannot be employed in the decomposition of Albert coal, nor any of the softer bitumens. These substances melt, and adhere to the iron closely, and therefore cannot be agitated like dry coals, when they are heated.

With the above-mentioned objects in view, namely, the agitation of the material while it is exposed to heat, oscillating retorts have been recommended and patented. Iron bars are fixed longitudinally in the cylinders, to prevent the charge from sliding, and to insure its rolling over.

BRICK OVENS.

Brick ovens have been introduced to decompose coals and produce oils. They are made of fire-brick, and laid in fire-clay. Their form is such, that the heat is distributed over a large surface. These ovens are incapable of resisting pressure, and they are apt to crack and grow leaky. If they are ever found to be economical, it will be in situations where coals and coal shales are cheap and plenty, and where the loss of vapor and fuel are not things of large account.

VERTICAL RETORTS.

In France, at Mehlam on the Rhine, and other places in Europe, upright retorts are used. They have been employed in Ireland for the distillation of peat. They are filled from above, and when the charge is exhausted it is drawn from beneath. They require a great deal of fuel. The yield of oil is small and impure.

Patents have been granted in the United States for several vertical retorts, in which improvements are supposed to have been made upon those used in the Old Country; but in none of these have the advantages sought for been obtained. It is obvious that the discharge of the gases from which the oils are condensed must take place above the mass of the material in the retort. The sooner the oily vapor of the charge is removed from the retort and condensed, the greater will be the amount of oil produced; for if that vapor is exposed to a heat equal to that by which it was first formed, it will itself be decomposed, and a part of it converted into permanent gases. Again, the vapor first formed will be deprived of a part of its hydrogen, and there will be a diminution in the quantity of lamp oil.

Agitators, or stirrers in retorts, have been introduced for the purposes before-mentioned. Count de Hompesch patented and used an Archimedean screw nearly twenty years ago. By means of this screw the material was agitated, and finally discharged at the end of his retort. Several American patents have been granted for machinery to stir or agitate the charge of material, both in horizontal and upright retorts during its distillation. In situations where coal is abundant the value of these inventions will be carefully weighed against the complexity of the machinery and its constant wear.

In order to apply a certain degree of heat to the substances undergoing distillation, baths of fusible metal have been placed in retorts and stills, the melting point of the metal being adjusted to the degree of heat required; but the experienced distiller calls for no such aid.

CLAY RETORTS.

Clay retorts were used in the manufacture of coal gas many years ago, and a contest has long been carried on between their advocates and those who prefer iron for that purpose. In Europe the clay retort is gradually coming into use. In Scotland the old iron cylinder is now seldom seen in gas works. The manufacturers of coal gas in the United States are yearly submitting clay to the test; but up to the present time, of all the retorts in operation, a very few are composed of that material. When the clay cylinder is first charged, gas escapes through the fine fissures opened by the baking of the substance. These openings, however, are soon closed with carbon, and the retort is perfect. The chief advantage of the clay retort is its durability. In the distillation of coals for the production of oils, they are, doubtless, valuable, and the ordinary mechanic of the country understands the methods by which they are put in working order. In their use, it should always be understood that they will not withstand as much pressure as iron, and therefore their discharge-pipes should be large, and their condensing apparatus open and free.

Among the numerous means applied to the extraction of oils from coals, coke furnaces merit some attention. The cutting and piling up of mounds of wood, covering them with earth, and firing them to obtain charcoal, is a process familiar to almost every one. In this operation all the volatile products of the wood escape in gas and smoke, and are lost. Within the past century charcoal furnaces have been invented by which those volatile products are collected, and the distillation of wood has produced a new class of substances; the chief of which are acetic acid, pyroxylic spirit, creasote, picamar, copnomor, paraffin, eupion, etc.

In China, Russia, and Sweden, the carbonization of wood is effected in pits, or furnaces not dissimilar to those for which patents have been granted in this country. The furnace is in the shape of an inverted cone, and the receptacle for the tar is at its side. Coal is converted into coke in a similar manner. In Europe coke has been extensively used in the manufacture of iron. In Great Britain it is burned on railways to avoid the smoke produced by coals, and coking furnaces are in constant use for its supply. In 1781 the Earl of Dundonald obtained oils by heating a quantity of coals in a coke furnace. The oils were condensed from the expelled vapors, and coke remained.

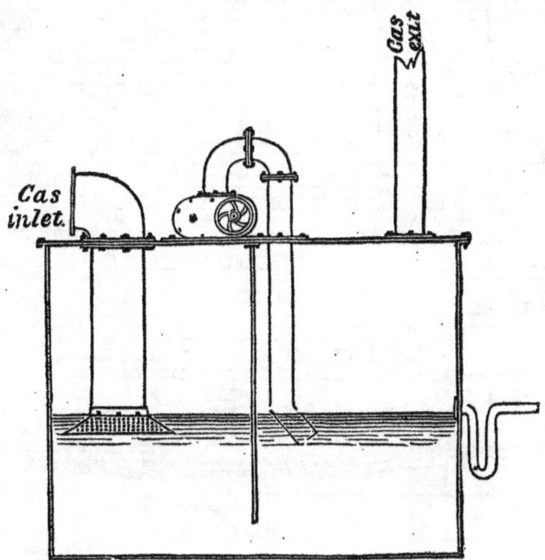

Exhaust and Condenser.—Section. Scale of Coke Oven.—Plan and Section.

The manufacture of coal gas now supplies vast quantities of coke, and the oils are called coal tar.

In August, 1853, two patents were granted in England for upright coking furnaces, the object being to obtain crude oils, and not coke. In these, and in other instances, the coals are produced in large perpendicular cones, or cylinders of masonry. A fire is lighted below, and as it advances upwards the volatile parts of the material are driven off by the heat produced by itself, and without the aid of any external heat. Discharge pipes are fixed at the top of the furnace, and communicate with a condenser in which the oils are formed.

The first objection that presents itself to this method of obtaining oils is the admission of air to the material, by which combustion rather than distillation is the result. To

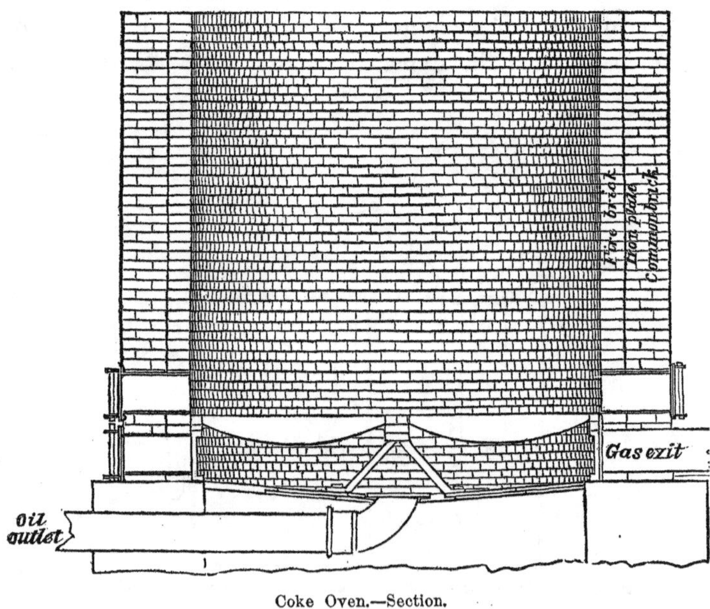

Coke Oven.—Section.

afford a remedy for this difficulty Mr. Little obtained an

English patent in 1854, the invention of which is to draw, or drive through the fire a blast of air, which is then said to be "*burned*," or deprived of its free oxygen, so that the combustion of the material is avoided, and the distillation carried on by the heat afforded by the gases emanating from the material itself. In this process the charge contained in the coking furnace is first fired at the bottom, then a current of air is drawn through the fire and the material in the furnace by an aspirator or exhausting pump, the oily vapors being drawn into condensing chambers, or worms, in the manner practised in ordinary distillations. An upward distillation has been opposed on the ground that the oil, which at first would be at the top of the fur-

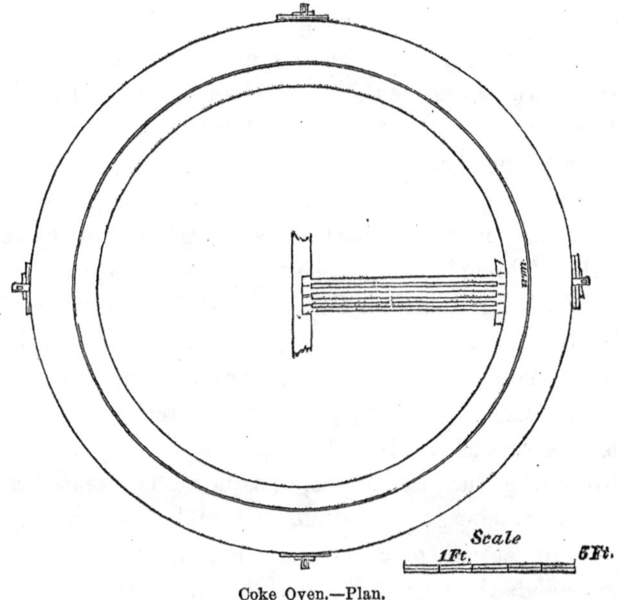

Coke Oven.—Plan.

nace, falls back, and undergoes repeated decompositions before its vapors finally escape. In practice this objection is groundless, for if the vapors from which the oils are condensed are light, they make their exit immediately; if they are heavy, and condense in the furnace, their oils are improved by further decomposition.

Patents have been granted in the United States for similar coke furnaces. In one of these the current of air is directed downwards through the fire, material, and furnace, by a jet of steam thrown into the discharging pipe below. After a wood fire has been kindled at the top of the furnace, and a stratum of hot coals is spread over the charge, a downward current of air is started, and continued until all the volatile matter is expelled from the material. It does not appear, however, that reversing the air current is a matter of any importance in the operation; the chief object being to force it through heated bodies, and thereby deprive it of a part of its oxygen before it reaches the charge. Ingenious as this method of distillation really is, its economy is doubtful, for sufficient heat cannot be applied to the charge in the furnace without the admission of oxygen, and that oxygen, when admitted, results in more or less actual combustion, which reduces the quantity of oils. This method has been extensively tested by the New York Kerosene Oil Company, who, according to their published reports, distilled sixty-eight and one-seventh gallons of merchantable oils and paraffin from one ton of Boghead coal. By the large D-shaped retort seventy gallons of such oils can be obtained.

From what has been stated, this question presents itself—What is the cheapest, most efficient, and economical retort for manufactories of coal oils? Perhaps foremost in the reply stands the large horizontal D-shaped retort—next the revolving retort, which for running the greatest quantity of oil in the shortest space of time stands unrivalled.

Next in importance to the form and the mode of applying heat to the retorts is the condenser, or cooling apparatus, in which the gases and vapors of the material assume the liquid form. It may be laid down as a general rule, that the sooner the lighter vapors generated in the retort are withdrawn from it and cooled, the greater will be the yield of oil. The discharge pipes of the retorts employed

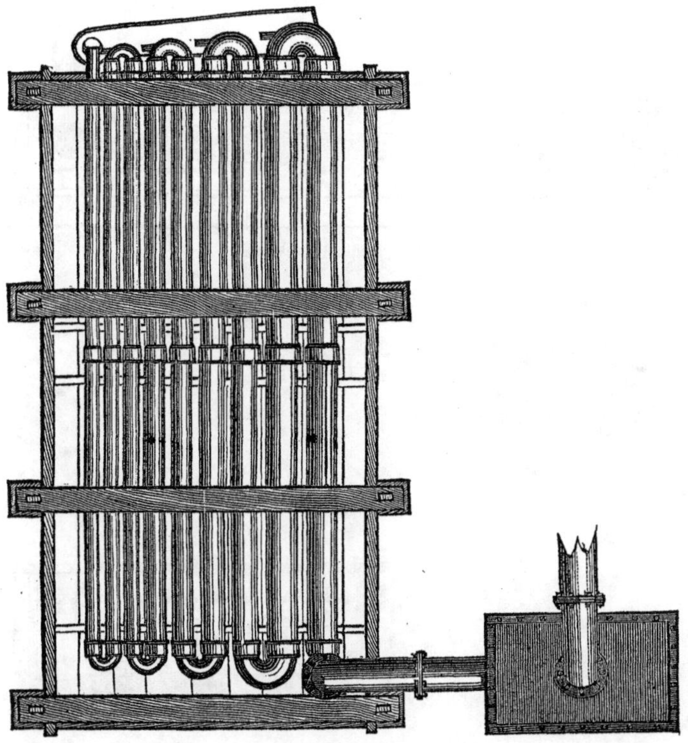

Crude Oil Condenser.—Plan.

in the manufacture of coal gas are upright cylinders, in which a part of the volatile products of the distilled coal are condensed, and fall back into the retort, where they are decomposed, and the quantity of gas thereby increased.

By this management the gas is made by the reduction of the hydrogen of the coal tar, which, consequently, contains

Crude Oil Condenser.

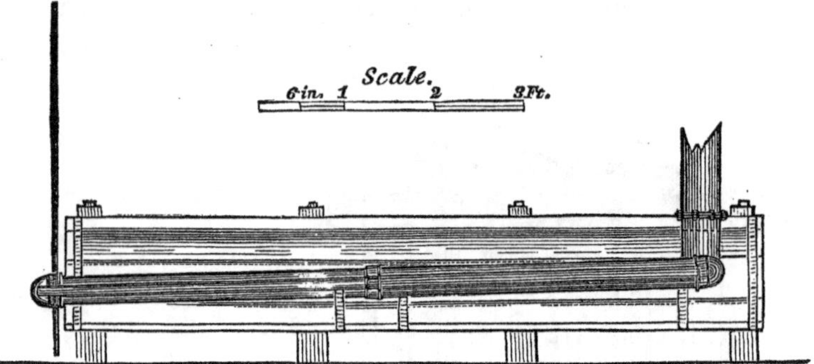

Longitudinal Section.

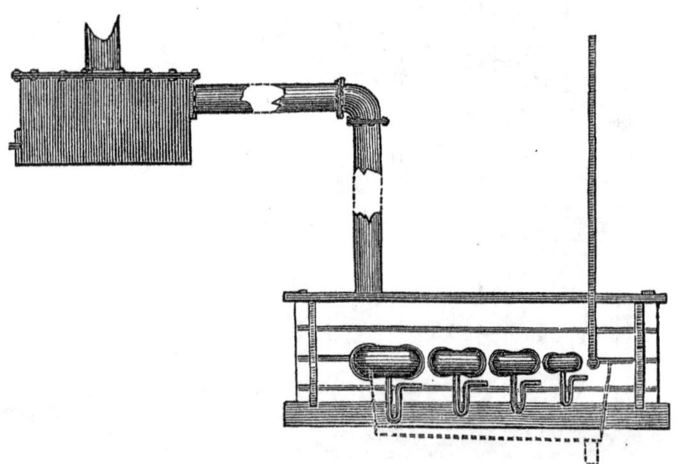

End Elevation.

much carbon, and is thereby rendered unfit for the manufacture of oils for lamps.* The exit, or discharge pipes,

* See table of homologous compounds.

should therefore open outwards from the retort, as near to the charge undergoing decomposition as may be convenient.

Again, pressure upon the vapors generated, and the retort itself, should be avoided as much as possible. The dipping of the discharge pipe in a main, to seal it against a return of gas, causes a pressure according to the extent of that dip. The greater the dip the greater the pressure, and the quantity of oil will be diminished accordingly. It is on this account that exhausting pumps have been applied to the gas pipe leading from the main. The effect is to exhaust the charge in one half of the time usually required for that purpose, and with less heat in the furnace. But exhausting pumps are expensive, and, when employed as above, require to be kept constantly in motion. Therefore when the crude oil is made at the mouth of a coal mine their economy will afford matter for consideration.

The condenser may consist of the common worm generally used in distilleries—a serpentine pipe passing through a cistern of water, or an open-chamber, all of which must be kept constantly cold by an influx of water.

But when much paraffin is present it is necessary to keep the water at a temperature of 70° or 80° to prevent the paraffin from cooling, and obstructing the apparatus. The gas that remains after the condensation has been completed may be collected in gasometers, and employed for illuminating purposes—to afford heat for the subsequent distillation of the oils, or to produce steam.

STILLS.

The variety of stills, and the contrivances applied to them, for the distillation of coal and other oils, equals that of retorts. Experience has led to the almost general adoption of cast iron stills for those purposes. They have been made with bottoms concave upwards and with hemispherical tops, with bottoms concave downwards and flat tops, some broad and flat, others high and cylindrical—some have been placed in steam jackets—many are exposed to the naked fire. The different opinions prevailing among the manufacturers prevent any settled form being established. Stills made of boiler-plate iron have been tried; but when a high heat is required, and they are exposed to the direct action of the fire, they are soon destroyed, or commence leaking at the rivets. When the heat exceeds 560°, which is necessary in the distillation of the heavier oils and paraffin, they are in danger unless they are protected by the admission of steam, and guarded against the fire of the furnace. Whether the still be of cast or sheet iron, it is always unsafe to run the oil down so as to "*coke*" its bottom.

In order to facilitate the flow of oils, stirrers have been placed in the still to agitate the charge during its distillation. The effect of these stirrers will ever be to render the distillation more or less imperfect, by lifting the impurities upwards into the current of vapor rushing outwards into the worm. Stills with double necks have been tried, but without any real advantage. Some have preferred a large still, and they have been made to contain three thousand gallons. Such stills are more liable to accident, and dangerous in the event of fracture than smaller ones, and have no superiority in regard to time in working.

The first distillation of the oils may be carried on continuously by admitting into the still a small stream after the heat is up and the distillate begins to flow from the worm. But this mode requires more than an ordinary degree of heat to compensate for the caloric taken by the inflowing oil to bring it up to the distilling point. Simplicity of machinery and steadiness of operation are always desirable; on this account, for reasons already stated, and from many actual trials, the author recommends that the

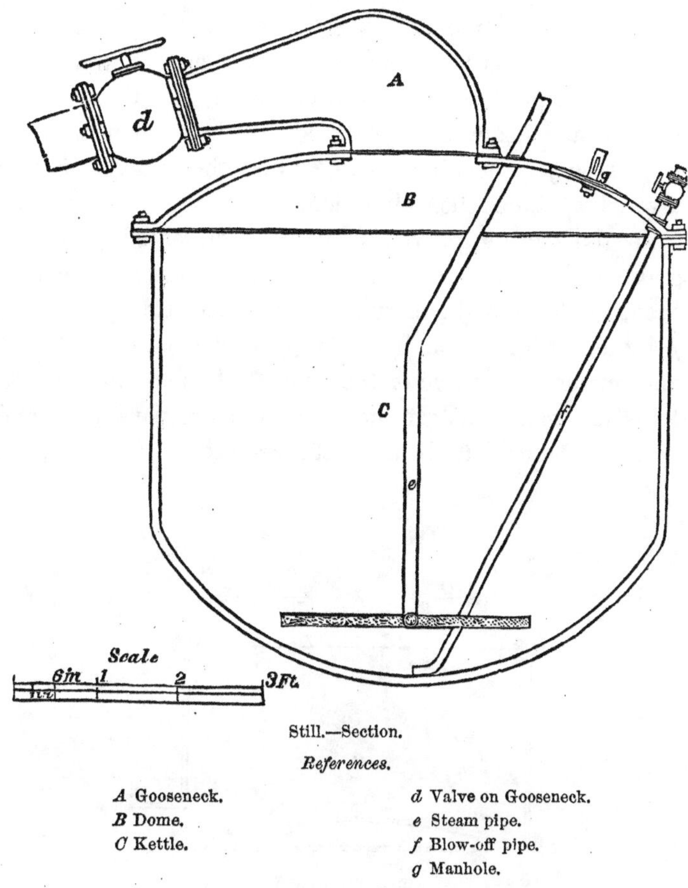

Still.—Section.

References.

A Gooseneck.
B Dome.
C Kettle.
d Valve on Gooseneck.
e Steam pipe.
f Blow-off pipe.
g Manhole.

largest stills, and those employed for distilling the crude oils as they come from the retorts or oil springs, shall not exceed in contents sixteen hundred gallons, and as there is generally a loss on the first distillation of ten or twelve per cent. in carbon and impurities, the working contents of the refining stills need not exceed fourteen hundred gallons. The diameter of the largest stills may be eight feet six inches, with a height of four feet six inches. The crown should be moderately concave upwards to the neck. Those stills must be carefully protected against the direct action of the fire by arches of fire-brick. Common or superheated steam may be introduced into them through large rose jets opening above, or into the charge. Steam always facilitates their operation.

In the cut on p. 75, a valve is represented upon the gooseneck. This has been used by some persons to enable them to blow out the tarry contents of the still by the pipe, *f*, by admitting steam by the pipe, *e*, closing the valve, *d*, and opening the valve or the pipe, *f*. When the still bottoms are made convex, however, the ordinary pipe and cock can be used to draw off the residuum. The draw-off

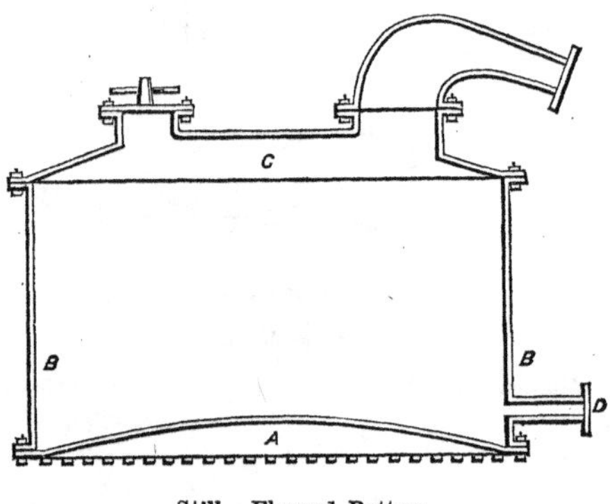

Still.—Flanged Bottom.

pipe may be of wrought iron of two inches diameter

inserted into the side of the still by a screw, and secured on its inner end by a locknut. The cock may be iron with a brass plug.

Where high temperatures are required, so as to distil various oils to a pitch or coke, the still represented by the cut, page 76, has been found to answer the purpose. The bottom, A, is two inches thick, and is flanged to the side, B, with three-quarter inch bolts, six inches apart. The sides are flanged to the dome or cover, C, in the same way. The sides form a cylinder when cast, and the outlet pipe, D, three inches in diameter and one foot long, is cast with it. This still, seven feet in diameter and four feet deep, was used by Henry Gesner in distilling candle tar, and by the author in distilling coal tar. In both cases the distillation was carried on until a coke remained, and the heat was so, that upon looking into the furnace the concave bottom of the still was of a bright red color. When the bottom was burned out, which was the case every two months, a new bottom was bolted to the sides.

The diagram on the following page exhibits an arrangement of still and worm, or condenser, in section. The still is shown with a loose bottom, but that is not absolutely necessary. The sides and bottom may form one casting.

The still, A, is a cylinder of cast iron seven feet in diameter and one inch in thickness, four feet deep, sitting in a groove, V, which is carried around the bottom, B. This bottom is one and a half inches thick. The dome, C, is low, not rising more than one foot five inches above the level of the top of the side. The gooseneck, D, is three quarters of an inch thick, one foot three inches wide, where it is fastened to the dome, and tapering to eight inches where it joins the pipe which connects it with the worm, E. The pipe connecting the gooseneck and worm tapers from

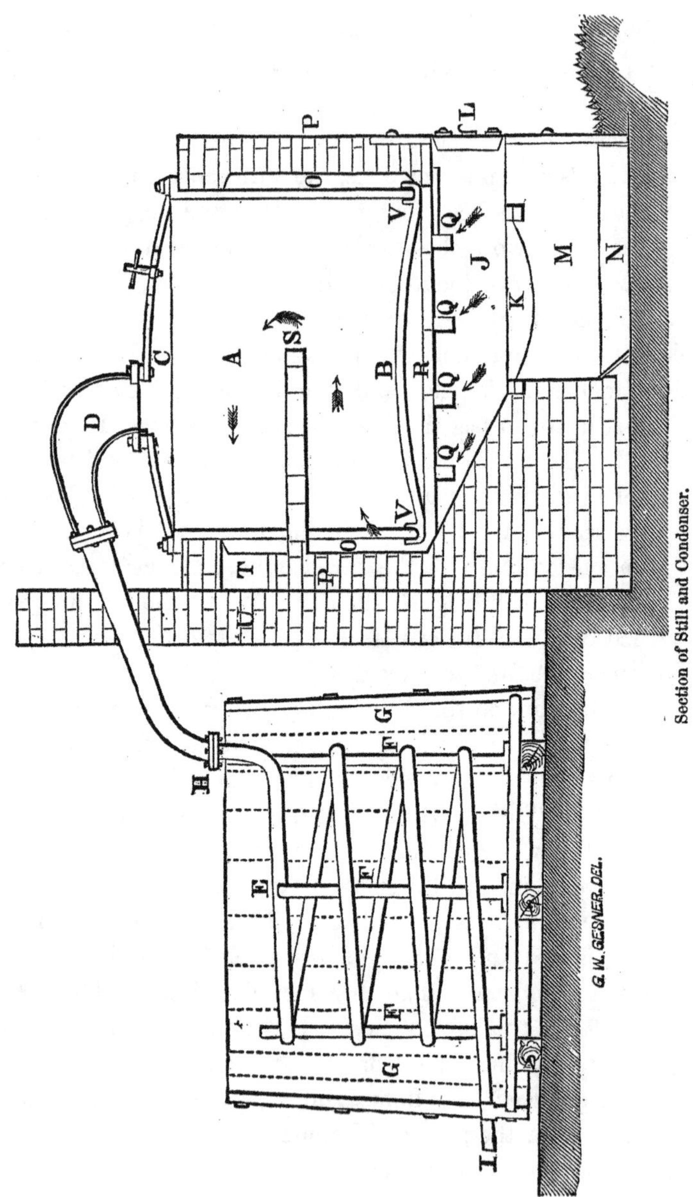

Section of Still and Condenser.

eight to four inches where it joins the worm, E. The worm is a spiral coil of wrought iron pipe, fastened securely by iron stays, F, into the worm tank, G. This tank is eight feet in diameter, and six feet deep. The worm coil contains one hundred feet of pipe, tapering from four inches in diameter at H, to two and a half inches at the tail-pipe, I. It is made by joining equal lengths of four inches, three inches, and two and a half inch pipe. Where the tail-pipe passes through the worm tank it is secured and leakage around the pipe prevented by two locknuts, one on the inside and the other on the outside of the stave, a thread being cut on the pipe long enough to pass through the stave and afford room for the locknuts to be applied. The worm tank is seven feet six inches in diameter and six feet deep, made of three inch staves, and bottom, and hooped with four hoops, three inches wide and three quarters of an inch thick.

The still is heated by the furnace, J, when open fire is used. The fire bars, K, are four feet in length, and cover one foot six inches in width. The furnace door, L, is one foot four inches high, and one foot three inches wide. The ashpit, M, corresponds with it in size. The water pan, N, corresponds with the ashpit in width, and is six inches deep. The length of ashpit and pan is the same as that of the furnace bars. The space, O, around the still is four inches wide, and the walls, P, are one brick or eight inches thick. The throttles, Q, are eight inches deep and four inches wide. They are small flues which distribute the heat around the still. The still bottom rests upon fire tiles which are laid in a circle to suit it. These tiles cover the throttles. The bridge, S, prevents the heat from escaping at once, by the flue, T, to the chimney, and brings it forward around the front of the still. The wall, U, is the division wall between the still house and refinery. (See drawings of refineries.)

In some factories the wall around the still is made in sections, so as to be easily removed when a new kettle or bottom is to be inserted. This plan is the most convenient, though the most expensive at the outset.

The still commonly employed by the American petroleum refiners, is shown in the diagram annexed. It is a

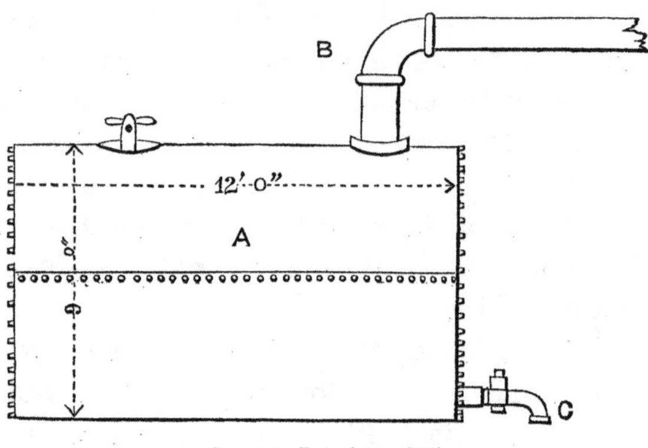

Common Petroleum Still.

cylinder of boiler iron twelve feet long and nine feet in diameter. Its working contents are eighty barrels, or three thousand two hundred gallons. The gooseneck is a four-inch elbow connected with four-inch wrought iron pipe. The draw-off cock is at the bottom and end furthest from the fire, which is applied much in the same way as to a steam boiler. The plates are in twelve feet lengths and present no riveted seams to the action of the fire. The use of wrought iron for stills is perhaps the most economical in the aggregate. Cast iron is exposed to the danger of fracture when it is cooled suddenly. A current of air from an open furnace door will often cause the bottoms to crack. In such case, the mode of repairing is by a plate of boiler iron being bolted down over the fracture.

There have been many alleged improvements in the

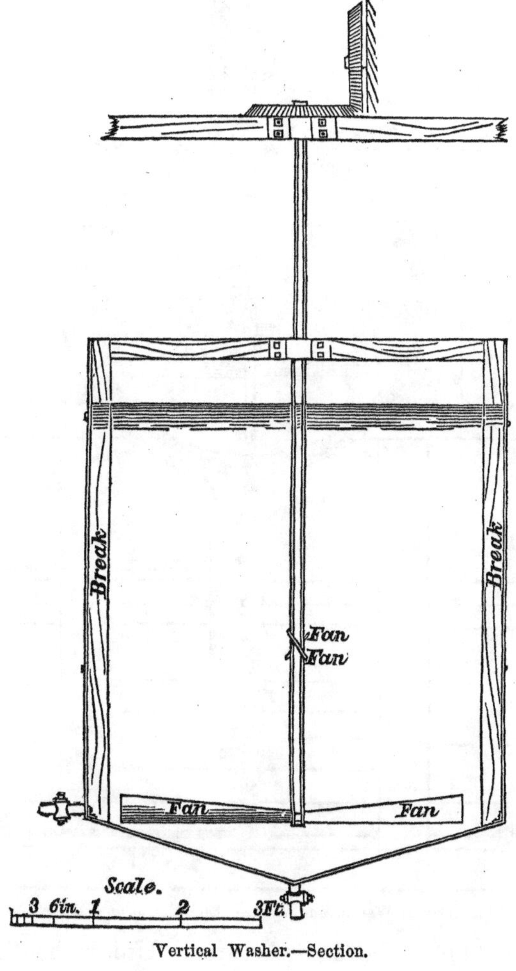

Vertical Washer.—Section.

construction of stills. Most of them involve either care,

which can hardly be expected from the ordinary workmen; or the use of expensive methods of obtaining heat. Simplicity in construction and in working must be the prime

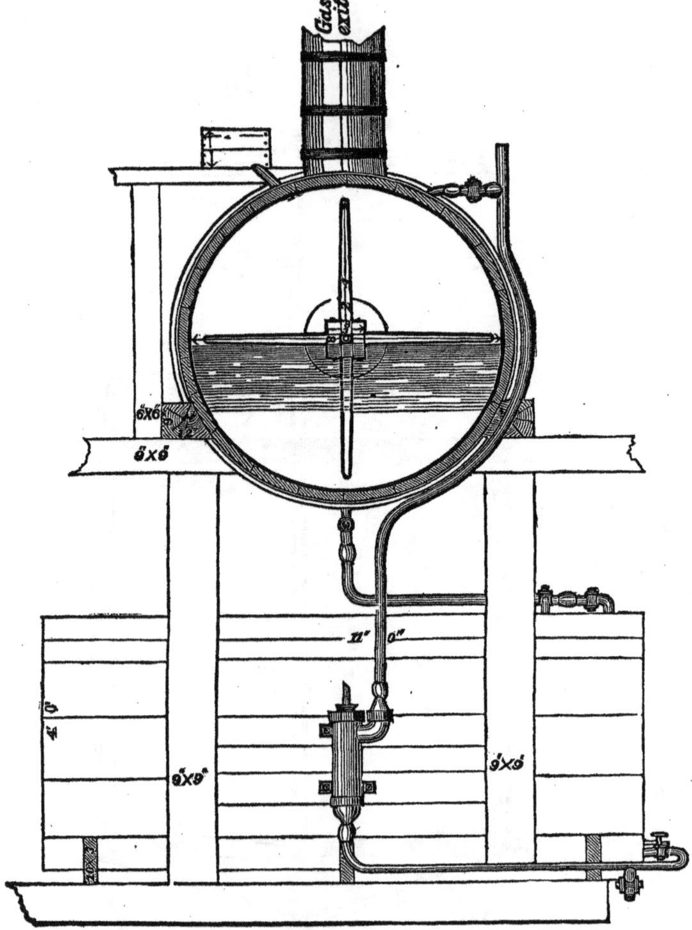

Horizontal Washer and Tanks.—End Elevation and Section.

consideration of the manufacturer. Unless the distillation is so regulated that every twenty-four hours produces a certain amount of oil, no correct idea of the profits of oil refining can be had; and the still, being most apt to be

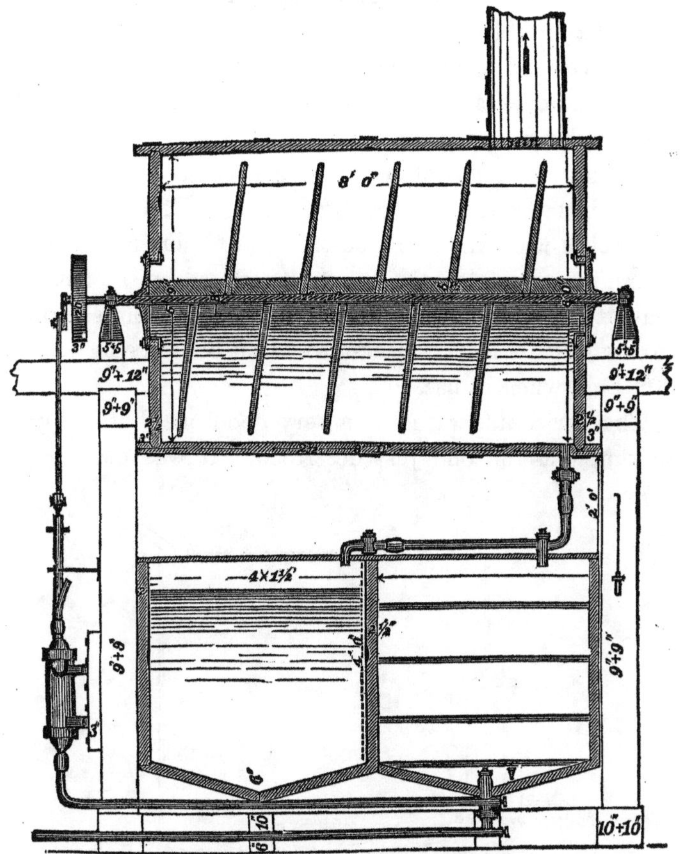

Horizontal Washer and Tanks.—Longitudinal Section.

out of order, should be of the most simple form, and admit of cheap and easy repair.

WASHERS OR AGITATORS.

These may be vertical as on page 81, or horizontal as on pages 82 and 83.

The vertical washer is of light boiler iron, having a concave bottom to assist the drawing off of the tarry matter thrown down by the reagents.

A shaft with fans of wood or iron, placed obliquely, as in a ship's screw, is turned by gearing. Breaks of wood, six inches wide and one inch thick, are secured to the side. These serve to break the current produced by the fans, and make the agitation more thorough. The bottom should be surrounded by a steam-jacket to admit of its being kept at 90° Fah. when in use.

The horizontal washer is a very good one, but not quite so simple as the one just described. It was used by the

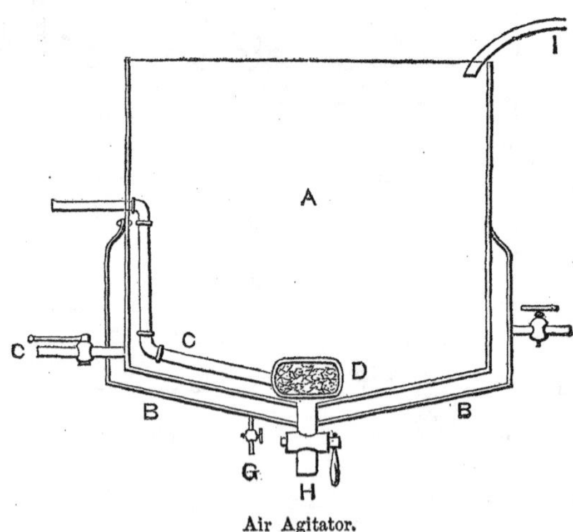

Air Agitator.

author in connexion with a system of tanks for treating coal tar products. The pump conveyed the oil to be treated to the horizontal chamber, which was lined with lead, and had a shaft running through it with arms or beaters of wood. The oil, after being agitated with acid, was let down into the tank below to be settled, while the agitator was set to work upon another charge.

An air agitator, shown on preceding page, has been found to be the most convenient and thorough. It consists of a vessel of thin boiler iron, A, surrounded at the bottom and part of the sides by a steam-jacket, B. A two-inch wrought iron pipe, C, communicates with a blowing fan, such as are used for small blast furnaces, or with an air-pump. The pipe, C, enters a perforated iron vessel, D, or may be left coiled with its end open on the bottom of the agitator. This arrangement leaves the interior of the agitator clear of all obstructions. A steam-pipe and valve admit steam to the jacket. The drip-cock, G, carries off the condensed steam. The cock, C, is the outlet for the oil after agitation, where a settling vessel is used, or the oil is discharged into another agitator as in coal oil refining. (See drawings, of Refineries.) The pipe, I, is the inlet from the pump. The cock, H, is for drawing off the acid residuum. The cut represents an agitator seven feet in diameter, and six feet deep.

SUPERHEATED STEAM APPARATUS.

Steam can be superheated for distilling purposes by any mode which involves its passing over hot surfaces before entering the charge of oil in the still. A bench of three ordinary gas retorts, connected with pipes, would form a very good superheater.

In the above diagram is an arrangement for superheating which has been found satisfactory. A series of cast iron pipes, B, four feet in length, two inches in diameter, metal two inches thick, slightly arched, and connected by return bends, D, are placed in an oven, A. The damper, L, regu-

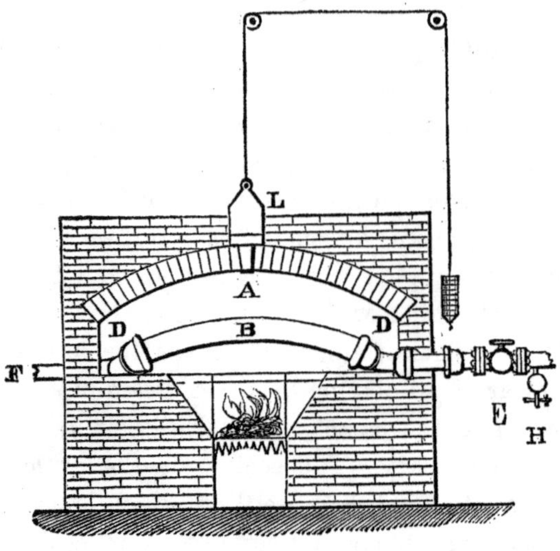

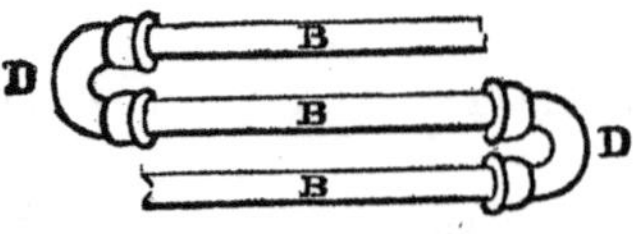

Superheated Steam Apparatus.

lates the heat. The fire is prevented from striking directly upon the pipe, by fire tiles placed under them, with small flues passing through them between the pipes. The inlet of steam is regulated by the valve, E. A small globe and drip cock, H, should be placed on the boiler side of the valve to insure the drawing off of any water before letting steam into the superheater. The pipe connecting with the boiler should enter a drum on the top of the boiler, so as to get the steam as dry as possible.

The exit from the superheater is a wrought iron pipe of the same diameter, carried into the still at the top and carried down its side, and coiled once or twice around its bottom. This pipe is perforated with holes three-sixteenths of an inch in diameter. The still may be set in sand. A pressure of forty pounds in the boiler is the proper one to work with. A mode of working the superheater is given further on.

Andrew McLean, of Liverpool, has been very successful

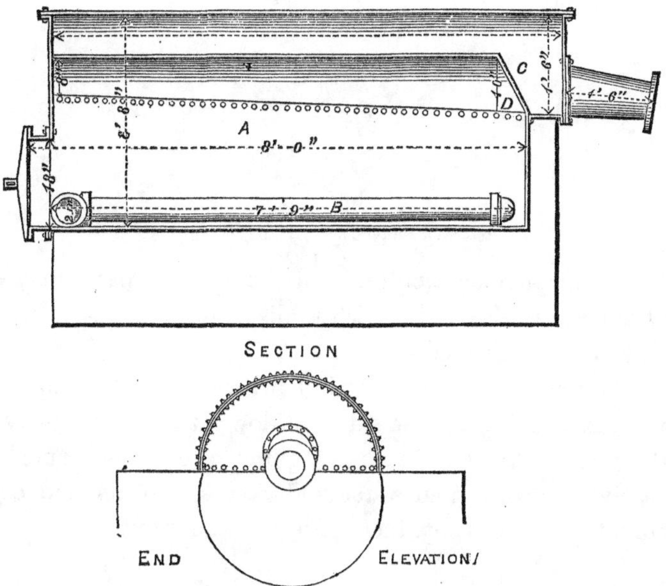

Retort for Bituminous Clays, Asphaltum, etc.

in introducing superheated steam in coal and petroleum

distillation. His application of superheated steam to the distillation of bituminous shales in Great Britain, is of great utility.

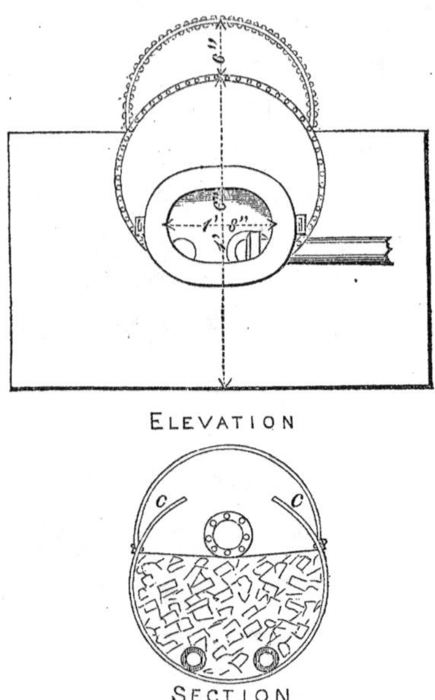

Retort for Bituminous Clays, Asphaltum, etc.

The arrangement on this and the opposite page, may be of service to those who wish to distil bituminous sand or clays by superheated steam.

The retort, A, is set in sand or brickwork. A perforated pipe, B, running up one side and down the other, conveys the steam to the charge; C, is a gutter or channel formed to receive the condensed water and vapors, and its end continuation, D, prevents their return to the retort.

CHAPTER V.

Products of the distillation of wood, coals, asphaltum, bitumen, petroleum, and other substances capable of yielding oils.

PRODUCTS OF THE DISTILLATION OF WOOD.

THE products of wood distilled in close vessels are very numerous. The resinous woods give results different from those not resinous, and each kind affords some peculiar products. During distillation all yield more or less carbonic acid, carbonic oxide, and carburetted hydrogen. Charcoal remains in the retort. Some of the products are soluble in water, others are not. Of the products soluble in water and volatile, there are *acetic acid, or pyroligneous acid.* This is the most abundant liquid. It contains much creasote, and preserves meat, giving it at the same time a smoky taste and odor.

Pyroxylic spirit.—By distilling the crude pyroligneous acid a mixed liquid is obtained, known as pyroxylic spirit, or hydrated oxide of methyle. From this spirit Gmelin and Liebig derived lignone, xylite, xylitic acid, naphtha, xylitic oil, and resin, mesetine, methol, mesite, acetone, and other volatile liquids have been obtained, of which, up to the present time, there is but an imperfect knowledge existing.

PRODUCTS OILY AND VOLATILE.

Among these creasote is predominant. This is a clear neutral oil, with an odor of smoke, and hot pungent taste.

It evaporates without residue, and is turned to a brown color by being exposed to the light. It is soluble in ether, alcohol, acetic acid, ammonia, and potash—is used as a styptic, and considered as a valuable remedy for the toothache. Creasote has also remarkable antiseptic properties, and is employed in dyeing and tanning. A distinction between creasote and carbolic acid has not been clearly made out.

Picamar was discovered by Reichenbach, with creasote in the heavy oil of tar. With potash, it forms a crystalline compound. It is a colorless oil, having a hot, bitter taste. Its composition has not been clearly described.

Copnomor.—With the creasote and picamar the above chemist discovered copnomor, a limpid, colorless oil, highly refractive, with an aromatic odor and styptic taste. Nitric acid converts it into oxalic acid, nitro-picric acid, and other complex substances, of which little is known.

Eupion, another oily, or rather spirituous liquid, discovered by Reichenbach in the oil of tar, is $C_5\ H_4$. It is readily purified by distillation, and has a specific gravity of 0·740. The author obtained it from the tar of candle manufactories, with a specific gravity of 0·640, and a boiling point of 112°. It is, therefore, among the lightest liquids known. It resists the action of the strongest sulphuric acid. With nitric acid it forms several new combinations analogous to those of benzole. It is perfectly colorless, evaporates rapidly, and to some persons it has an agreeable odor. This oil does not exist ready formed in the tars, but is produced by the action of strong acids and alkalies upon the distillates of crude oils. In the manufacture of hydro-carbon oils, eupion includes a number of the members of the homologous compounds of carbon and hydrogen. It is now frequently sold as benzole, and employed for making what is called the benzole or atmospheric light, and for removing oil stains from clothes. A number of liquids have been classed under the denomination of eupion; they are all hydro-carbons, and their formula is C, H, or C_5 H_4, or some multiple of it.

Eupion may not only be distilled from wood, but also from other substances capable of yielding tars by distillation. It burns with a brilliant white flame, free from smoke; but it is extremely inflammable, and a dangerous liquid for lamps.

SOLID PRODUCTS OBTAINED FROM THE DISTILLATION OF WOOD.

Paraffin is the name of a white solid substance, or silvery scales resembling wax, discovered by Reichenbach. It is formed in large quantities from the petroleum of Rangoon, and the author has obtained it from the Ouachita coal of Arkansas, at the rate of 143 lbs. per ton. Coals, asphaltums, bitumens, petroleums, peat, and other substances, afford paraffin from one to five per cent. of their oils. It is most abundantly produced by the distillation of wax with lime.

Paraffin melts between 110° and 114°. Its specific gravity is 0·870, and according to Lewis its formula is C_{20} H_{21}. It is readily made into candles, and in a wick it burns with a beautiful, clear, white light; and the candles are semi-transparent. It is indifferent to the strongest acids and alkalies. A number of compounds of carbon and hydrogen have been confounded with paraffin, such as methylene, ethylene, butylene, etc. It is remarkable that the paraffin produced by the distillation of different kinds of materials differs considerably on some points of comparison, some having a higher, and some a lower melting point. These differences, however, may arise in some degree from the amount of heat by which they are produced, and their treatment to render them pure. The greatest obstacle to the application of paraffin for candles is its low melting point. It may be mixed with bleached wax, which does not fuse, in general, below 154°. The cup-like cavity around the wick of a pure paraffin candle is apt to yield to the heat, and the melted material overflows, and bears with it the name of "*slut.*" Doubtless there are improvements to be made in the manufacture of this beautiful article. Paraffin does not exist in coal ready formed. It is one of the combinations resulting from the interchanges of the elements of bituminous and other bodies during their exposure to a high temperature. Paraffin burns well in the kerosene or common coal-oil lamp, when dissolved in hydro-carbon oils; but in cold weather it hardens, and will not then ascend the wick.

Cedriret is a volatile solid which forms red crystals in a solution of sulphate of iron. These crystals dissolve in sulphuric acid, and the red color is changed to blue. The blue tinge produced by reflected light of some of the coal oils in the market owes its origin in part to the presence of cedriret.

Pittical.—When heavy oil of tar is neutralized by potash, and barytic water is added, the solution is of a deep blue color, from the presence of pittical, which, when pure, is like indigo. It color has been fixed on cloth, but its manufacture has not yet been brought to perfection.

Pyroxanthine is another volatile crystalline solid, first obtained by Scanlan from pyroligneous spirit. Its crystals are of a fine yellow color, easily fusible. Its composition is represented to be C_{21} H_9 O_4.

The foregoing are the principal products of the distillates of wood. Besides these, there are others which are constantly engaging the investigations of chemists. They are important, and in time they will probably be applied to useful purposes. When the different kinds of wood, the different chemical changes produced by different degrees of heat, and the variable operations of re-agents are considered, it is not surprising that this division of chemical science should advance so slowly, and so little should be known of the changes matter undergoes by seemingly invisible agents. The identity of most of the before-mentioned products, with those resulting from the distillation of coals, affords much additional evidence that coal and bitumen, like wood and turpentine, have had one common origin.

PRODUCTS OF THE DISTILLATION OF COALS AT A HIGH HEAT, OR COAL TAR.

Certain specific spirits and oils have been obtained by chemists from coals and other bituminous bodies. These spirits and oils have been distinguished one from the other by their densities, boiling points, and other characters, and have received different and sometimes very inappropriate names. From coal tar Peckston distilled oil of tar and spirits of tar. Laurent, Reichenbach, Hoffman, and others, have given the composition of coal tar. Wagenman applied himself to the oils derivable from turf, brown coal, and bituminous slate, from which he obtained photogen, solar oil, and paraffin. From the slate near Bielefeld, Engelbach distilled light oil, heavy oil, butyric fat, and asphaltic fat.

Mansfield in his patent, registered in 1847, describes

alliole, benzole, tuluole, cumole, cymole, and mortuole, products collected by him from the distillation of coal tar. Among the oily substances obtained by the distillation of coal tar the following have been described:*—

Benzole	$C_{12} H_6$
Cumene, or cumole	$C_{18} H_{12}$
Toluole, or toluene	$C_{14} H_8$
Naphthalin	$C_{20} H_8$
Anthracene, or paranaphthalin	$C_{30} H_{12}$
Chrysene	$C_{12} H_4$
Pyrene	$C_{10} H_2$
Ampaline.	

ACIDS.

Carbolic	$C_{12} H_5$ O. H. O.
Rosalic.	
Brunolic.	

BASES.

Ammonia	N. H_3
Picoline, or odorine	$C_{12} H_7$ N.
Aniline	$C_{12} H_7$ N.
Leucoline, or quinoline	$C_{18} H_7$ N.
Parvoline	$C_{18} H_{13}$ N.
Lutidine	$C_9 H_9$ N.

and others not yet fully investigated.

Besides the foregoing compounds, derived from coal tar, phenyle, pyrrole, animine, olanine, cyanole, benzidam, etc., and others have been described. It has been usual to separate the coal tar of gas works into two parts, namely, naphtha and *dead oil.* The tar itself always contains much finely divided carbon, the quantity of which is augmented by a high heat. Both the naphtha and *dead oil* consist of a number of hydro-carbons. These cannot be considered

* See Gerhardt, *Chem. Organ.* vol. iv., p. 426.

as certain compounds, as they are liable to great variations. The nature of the coal, and the heat applied, as before remarked, have much to do with the quality of the tar, cannel coal being always more productive of spirits and oils than common bituminous coal. Besides these various and variable products, several of them, if not all, have many derivatives, formed by their combinations with other substances. For instance, by the action of chlorine on naphthalin, we have, according to the nomenclature of Laurent, chlonap*tase*, chlonap*tese*, chlonap*tise*, etc. By the action of bromine, bronap*tase*, bronap*tese*, bronap*tise*, etc. The derivatives of aniline are represented as chloranoline, dichloranoline, trichloranoline, bromanaline, dibromanaline, tribomanaline, nitrodibromanaline, etc., and thus pages might be filled with the names of these uncertain combinations, a systematic arrangement of which has not been completed. These discoveries mark the progress of chemical inquiry, although they have not, so far, added much to manufacturing or commercial interests.

When coal tar, and especially that obtained from cannel coal, is submitted to heat in a still connected with a proper condensing apparatus, it is resolved into water, benzole, naphtha, and various heavy hydro-carbonaceous oils—charcoal remaining in the still if the distillation has been carried on to dryness. In the meantime decomposition has taken place, and products present themselves that did not exist in the undistilled material. Of these products a part is volatile, and another and the largest part is dense and not volatile. The former may be advantageously distilled over by the aid of steam at the temperature of 212°, the latter by superheated steam. Manufactories have been established where the coal tar is distilled down to a thick pitch, which is applied for roofing buildings; the dense oils are employed for fuel in glass works; the benzole and naphtha, after being rectified, are sold for dissolving gutta percha and india rubber, for varnish, and for producing the benzole light. The heavy oils abound in naphthalin, which has not yet been extensively applied to any useful purpose. The last of the distillate frequently contains paraffin oil and paraffin.

Among the valuable derivatives of coal tar is picric acid, Welter's bitter, carbozotic acid, or nitro-phenesic acid of some chemists. This acid was discovered by M. Guinon, of Lyons, and its composition is stated to be C_{12} H_3 N. O_4 O_2. This substance is obtained by acting upon coal tar, or coal tar naphtha, with strong nitric acid. It produces a beautiful yellow color, which is capable of being fixed on silks and woollen cloth. It is used in France and England as a dye. The yellow stain communicated to the skin by nitric acid, and which cannot be removed by washing, arises from the production of picric acid. Aniline also is converted into a violet-colored powder, which has been sold for $250 per lb., on account of the beautiful red and purple dyes it communicates to silks. Its colors are permanent, and exceed in delicacy any before discovered.

Benzole (C_{12} H_6). *Bicarburet of hydrogen* (Faraday). *Benzine* (Mitscherlich).—This oil, so called, although it is rather a spirit, was discovered by Faraday, and by him condensed from oil gas. Mitscherlich obtained it by distilling benzoic acid with hydrate of lime, and it may be procured by passing the vapor of benzoic acid through a red hot tube. It exists in considerable quantities in coal tar naphtha, from which it may be separated by fractional distillation. It is readily purified, by first washing it with sulphuric acid, then with a solution of caustic potash, or soda, and final distillation over lime. Its specific gravity is 0·850, of its vapor 2·742, and it boils at 186°. Like other liquids distilled from coal tar, it is scarcely a distinct and separate product; but forms a member of a series to be noticed hereafter. Benzole holds a medium position between alliole, so called, and naphtha. With chlorine it forms chlorobenzole $C_{12} H_6 Cl_6$. Similar compounds are also formed with bromine, nitric and sulphuric acids. It is itself a starting point or type of a series of homologous compounds, the common difference at each step being $C_2 H_2$. These compounds all admit of their hydrogen being replaced by one, two, or three equivalents of chlorine, bromine, nitric acid, and amide; finally they give rise to bases, of which aniline or phenylamine is the type. *

It will be readily perceived how benzole differs from eupion. In both, the multiple, or increasing number of the hydrogen, is two; but as the benzole series starts with two equivalents of carbon to one of hydrogen and eupion, with one equivalent less of carbon than of hydrogen, the former series contains the most carbon throughout. In making the benzole, or atmospheric light,† the benzole requires to be diluted with alcohol, to prevent the flame from smoking. Again, eupion alone is found to be deficient in carbon for that purpose. A mixture may be made of the two liquids, in which the quantities of carbon and

---

* Gregory, *Outlines of Organic Chemistry*. London, 1852. 3d ed. p. 128.

† The benzole, or atmospheric light, is made by passing a current of air through benzole, or other volatile liquid hydrocarbon. The air, by taking up a quantity of the liquid, burns freely, and is distributed in the manner of coal gas. Numerous machines have been invented for forcing a current of air through the fluid, and some of them are very efficient. But below a certain temperature the air will not convey vapor sufficient to afford a good light. In cold weather, also, the vapor of the benzole condenses in the pipes, and the liquid itself requires the application of heat. These difficulties have so far been insurmountable.

hydrogen may be so adjusted that the light will be brilliant and without smoke.

By adding benzole gradually to strong nitric acid, with the aid of a gentle heat, a compound is formed which dissolves in the acid, and, on cooling, collects on the surface. On diluting the mixture with water, nitro-benzole is precipitated in the form of a yellow oil. This oil has a sweet taste, and the odor of the oil of bitter almonds. It is used in perfumery, and in the bakery. Benzole is employed for many useful purposes. It dissolves the gums, resins, and all fatty substances. It removes from cloth and silks spots of tar, grease, turpentine, etc., and for those purposes it has been imported from France in small bottles, which are sold at high prices. Its rapid evaporation renders it also a substitute for alcohol and turpentine in the preparation of paints and varnishes.

Cumole (C_{18} H_{12}), when treated like benzole, its homologues yield a crystalline solid, which is fusible and volatile.

Toluole (C_{14} H_8) is another oil, analogous to and homologous with benzole. It boils at 226°, and has a specific gravity of 0·870. When treated with nitric acid it yields two compounds, nitrotoluole and dinitrotoluole. Deville obtained a series of compounds from toluole, in which the hydrogen was replaced by chlorine.

Naphthalin.—This interesting and remarkable hydrocarbon exists in almost all kinds of tar. In coal tar it is very abundant. It does not exist ready-formed in coal, but results from a high heat in its distillation, and an interchange of elements during the decomposition of the bituminous mineral. Creasote, or carbolic acid, is its usual companion, and seems to add to its quantity. By the repeated distillations of coal tar, naphthalin will crystallize at the bottom of the receiving vessel, and may be separated from the oils that accompany it by simple draining and pressure. It is rendered pure by agitation first with sulphuric acid, then with a strong solution of caustic soda or potash, and final distillation and crystallization. When pure, naphthalin is colorless, and forms beautiful flat and needle-shaped crystals; it evaporates rapidly, like camphor, and gives out a peculiar odor, unpleasant to some persons, but agreeable to others. Its taste is hot and pungent, and it corrodes the skin. A soap made from it was considered beneficial to the complexion. It distils with water, and, like camphor, sublimes and crystallizes against the sides of the bottle in which it is contained, and opposite the light.

Chlorine and bromine combine with naphthalin, and lay the foundation of a great number of compounds, which are formed by the substitution of the chlorine and bromine for hydrogen. The labors of Laurent have been successfully applied to this inquiry, by which a new field of research has been opened, and the doctrine of substitution more clearly established. Sulphuric acid exerts itself upon naphthalin, forming *hyposulphonaphthalic*, *hyposulphonaphthic* acids, etc. Thus, also, with nitric acid; but the number of these combinations, and the great length of their names, render full descriptions of them unnecessary in a work intended to be practical. Naphthalin is worthy of a trial in medicine, and may hereafter prove itself to be valuable in the arts. In its unpurified state it adds to the offensive odor of the oils distilled from coals, and increases the cost of their treatment.

Paraffin has been already described under the solid products obtained from the distillation of wood. Its yield from the coal tar of cannel coals is seldom more than one-fourth per cent. of the tar, and it succeeds the naphthalin in the distillation.

Anthracene, or *paranaphthalin* (C_{30} H_{12}), is polymeric with naphthalin, and is obtained from the heavy distillates of coal tar. It melts at 350°, distils at 398°, and crystallizes in thin, foliated plates. Like naphthalin, it is acted upon by nitric acid, which produces a series of compounds, oxygen taking the place of hydrogen. Thus we have hyponitrate of *ancethracenase*, bi-hyponitrate of *anthracenese*, etc.

Chrysene also is found to exist in the last divisions of the distillates of coal tar. It is a crystalline solid, of a yellow color, melting at 455°, and not soluble in many liquids.

Pyrene (C_{10} H_2) occurs with chrysene. It is acted upon by nitric acid, which produces a number of derivatives. Chemistry is mainly indebted to Laurent for the discovery and description of many of these combinations.

VOLATILE BASES IN COAL TAR.

Carbolic acid.—This is a colorless oil, which, in its general character, resembles creasote; and by some it is believed to be only a modification of that compound. It also occurs in the heavy distillates of coal tar, and boils at 380°. Like creasote, it is very poisonous, and may be used as a remedy for toothache. If a piece of pine-wood be dipped in carbolic acid, and then in nitric acid, it will become blue, which finally changes into brown. This acid has an offensive odor, which it imparts to coal oils, and thereby increases the cost of their purification.

Picoline (C_{12} H_7) is a volatile, oily base, discovered in coal tar by Dr. Anderson. It boils at 272°, does not discolor pine-wood, and is probably the *odorine* of Unverdorben.

Aniline has been termed *crystalline*, *cyanol*, *benzidam*, *phenylamine*, *phenamine*, *phenamide*, etc. This base occurs among the products distilled from coals, and those produced by the destructive distillation of animal matter. It is also described as having been obtained from indigo. The author found aniline an abundant product in the tar of stearine manufactories, and the oils distilled from shales, which contain the remains of fishes and *crustacea*. Aniline is a highly refractive, colorless oil, of specific gravity 1·020. When pure, it has a hot, pungent taste, and pleasant smell. It does not act on turmeric, but turns purple to green. With bleaching powder, it produces a purple color. This color is frequently seen in the coal oils of the market.

Leucoline, or *quinoline*.—This base is found to exist among the last and least volatile products of coal tar. It boils at 460°, has a disagreeable smell, and neutralizes acids.

Lutidine is another of these bases, the nature of which has been but imperfectly made out.

COAL TAR NAPHTHA, BENZOLE (OR BENZINE), NITRO-BENZOLE, ANILINE, AND ANILINE DYES.

Coal tar from gas works distilled over fire with one-fifth its weight of water produces naphtha, which is that portion of the distillate coming over with the water. Distilled by steam passed through the charge, coal tar yields a larger quantity of naphtha, but of poor quality for aniline purposes.

The crude naphtha is thoroughly agitated with three per cent., or three gallons to the hundred, of sulphuric acid at 66°, and permitted to settle for three hours. The naphtha is then drawn off and again washed with five per cent. of sulphuric acid, and settled for five hours. The naphtha is then drawn off and agitated with water in large quantity, to remove as much of the acid as possible. The water is drawn off. The naphtha is then agitated with ten per cent. of strong solution of soda-ash, together with three per cent. of milk of lime. Draw off the naphtha as before. Pump into an iron still and pass steam through the charge, collecting all the naphtha that comes over with the water. The naphtha should then be run into a tank under ground, and left for twelve hours closely excluded from the light, until the water, mechanically combined with the naphtha, has had time to separate, which usually takes twelve hours. The article is now rectified naphtha, and should stand the sunlight without change.

From this rectified naphtha benzole is obtained by distillation, in a still so constructed as to prevent the passing over of all fluids which require a heat greater than 212° Fah. to volatilize them. This can usually be managed by surrounding the still head with water. As this boils at 212°, it will keep the head at about that temperature and serve to show the degree of heat at that point.

Pump the rectified naphtha into the above still and distil *over fire*. The vapors and oils go over as follows:

1st. Alliole 2d. Benzole 3d. Toluole	These distil over at 212°.
4th. Cumole 5th. Cymole	These will not pass over unless the temperature is above 212°.

The alliole and toluole may be separated, in a great measure, from the benzole by another distillation, in which the first and last portions of the distillate are rejected and the middle portion taken. This middle portion is ordinary commercial coal tar benzole. If required still purer, it must be treated with one half pound sulphuric acid to each gallon of the benzole, and then be, after settling, well washed with water. Then treat it with a solution of one ounce nitrate of soda to the gallon, and wash clean with water.

Pure benzole boils at 176° Fah., and is entirely volatilized at 212° Fah.

NITROBENZOLE.

The formation of this article depending upon the action of nitric acid upon benzole, various modes have been employed to obtain it.

It can be prepared by adding slowly and carefully fuming nitric acid to benzole, assisting the reaction by a moderate heat. This operation can be made in a glass vessel. When the nitric acid will dissolve no more benzole, which is known by the ceasing of effervescence, the mixture is cooled by being placed in a water bath, when the nitrobenzole separates as an oily liquid. This is washed with water, and afterwards with a solution of carbonate of soda, and can be purified by distillation. It is a yellowish fluid, getting deeper color by exposure to the air, and has the odor of bitter almonds. Its specific gravity is 1·080. It boils at 213°, and crystallizes in needles at a low temperature. It is also prepared by permitting a small stream of the benzole and nitric acid to pass through a long worm well cooled, the nitrobenzole being collected at the lower end of the worm.

ANILINE.

One of the best and cheapest modes of obtaining aniline

from nitrobenzole is that of M. Béchamp. A mixture of iron filings two parts and acetic acid one part, with about an equal volume of nitrobenzole, is distilled, the reaction being assisted by a gentle heat whenever the effervescence ceases. Aniline and water are found in the receiver, the aniline being separated from the water by the addition of a very little ether, which, dissolving in the aniline, causes it to rise to the surface, when it is easily decanted.

A very spacious glass or earthen retort must be employed as the mass swells up violently, and it must on the small scale be connected with the receiver by a Liebig's condenser, and on the large by an ordinary worm and cooling tub, which must be well supplied with water.

Aniline is also prepared from an alcoholic solution of nitrobenzole, which, after saturation with ammonia, is heated with sulphuretted hydrogen until a precipitate of sulphur takes place. The liquor is then removed and repeatedly saturated with sulphuretted hydrogen until no more sulphur separates, assisting the operation by occasionally heating and distilling the liquor; an excess of acid is then added, and the liquid filtered. The alcohol and unsettled nitrobenzole are removed by boiling, and the residuum is distilled with caustic potassa in excess. The aniline in the receiver may be purified by forming it into oxalate of aniline, crystallizing the salt from alcohol, and distillation as before with caustic potassa.

Pure aniline is a colorless liquid, strongly aromatic, and burning to the taste. It is very soluble in alcohol or ether, and slightly so in water. It becomes yellow on exposure to the air, and distils at 200°. It cannot be frozen.

Aniline is prepared also from the heavier oils of coal tar by agitating them for some time with hydrochloric acid in excess in a glass globe on a small scale, or on the large scale in a suitable vessel of lead or enamelled iron. The clear portion of the liquid containing the hydrochlorates of the bases present is evaporated over an open fire until acid fumes begin to rise, when it is decanted and filtered. The clear filtrate is then mixed with milk of lime or with potash in excess, by which the bases, chiefly "aniline" and "chinoline," are set free under the form of a brownish oil. The whole mixture is then distilled, the portion passing over at 360° Fah., being collected separately. This is crude aniline, and is purified by re-distillation and re-collection at the same temperature, and by fresh treatment with hydrochloric acid, and careful distillation with excess of potash and milk of lime as before.

ANILINE DYES.

The details of the various processes by which aniline is made to perform a useful part in dyeing would be too voluminous for this work. The following extracts from the patents of Georges de Laire and Charles Girard will serve to give an idea of the manipulation for aniline red. This patent bears date November 22d, 1860:

"By our new process we put into a distilling apparatus twelve parts arsenic acid and twelve parts water, and the arsenic acid having become completely hydrated, we add ten parts kyanol (the 'aniline' of French chemists). The whole is then agitated or shaken, so as to produce a thorough mixture, forming a homogeneous, clammy, or nearly solid mass. This mass is heated at a low fire, so as to gradually raise its temperature, when the liquefaction takes place. By this *modus operandi* water, and only a small quantity of kyanol, are evaporated, if the operation is conducted with proper care. At a temperature of 248° Fah., a great quantity of the kyanol or aniline is already converted into coloring matter, and care should be taken to keep the temperature at that point for some time, after which it is further raised; but in no instance above 320° Fah. We thus obtain a perfectly homogeneous and fluid mass above 212° Fah. This operation lasts from four to five hours. When cooled, the mass solidifies and becomes a hard, brittle matter, of a coppery hue, similar to Florentine bronze. This matter is highly soluble in water and other solvents, such as alcohol, and imparts to them a fine pure red tint, having no admixture of violet, the intensity of this coloring matter being so great that after having been boiled and concentrated it appears altogether black.

"This coloring matter may, without inconvenience, be directly applied to dyeing or otherwise coloring fabrics and other substances, the substance thus colored not retaining the slightest trace of arsenic. The arsenic may also be eliminated in an easy way by either of the following processes:—First process. The mass is pulverized and digested with either chlorhydric (hydro-chloric) or sulphuric acid, diluted with water. The clear solution thus obtained is then saturated with a slight excess of soda or carbonate of soda; thus the coloring matter precipitates, while the arsenic is dissolved in the alkali. The coloring matter is next washed once or twice in cold water, when it may be filtered or decanted.

"Second process. The coloring matter, after having been dissolved in water, is digested with a quantity of lime corresponding with the portion of arsenical compounds contained in it, the lime being slightly in excess. The coloring matter is then precipitated, as well as the arsenical compounds, which combine into insoluble calcareous salts.

The precipitates both, and the solutions, are then (without being separated) acted upon by either of the carbonic, tartaric, or acetic acids, which dissolve the coloring matter, the whole of the arsenic remaining insoluble."

Messrs. de Laire and Girard also obtain a violet coloring matter, and also a combination of blue and other coloring matter, by merely changing the proportion of the arsenic used. By acting upon ten parts of aniline, or any salt containing ten parts of aniline, with eighteen, twenty, and twenty-four parts of arsenic acid, they obtain a more or less violet, tending towards the pure blue.

Hofmann procured aniline red by the action of bichloride of carbon on aniline. M. Verguin prepared the color by tetra-chloride of tin. MM. Renard Brothers in France, and Messrs. Simpson, Maule, and Nicholson in England, manufacture the various tints which, under the name of Fuchsine, Magenta, Solferino, Mauve, and many others, are now so well known to the world.

Aniline green was produced by the action of aldehyde or wood spirit, upon the aniline red by M. Eusebe. It may be prepared as follows:—150 grammes of sulphate of rozaniline are dissolved in 450 grammes of cold, diluted sulphuric acid (three parts of acid to one of water). When the solution is complete, 225 grammes of aldehyde are added, the mixture being stirred. The whole is now heated in a water bath. From time to time a drop of the mixture is taken up with a stirring-rod, and dropped into slightly acidulated water, and as soon as a deep green solution is obtained the reaction is stopped. The mixture is now poured into thirty litres of boiling water, and to this solution are gradually added 450 grammes of hyposulphite of soda, dissolved in the smallest possible quantity of water. The whole is now boiled for some minutes. All the green remains in the solution, which may be used to dye silk.*

Aniline yellow is produced by the action of hydrated antimonic or a stannic acid upon aniline.

Aniline has been studied by Runge, Zinin, Futzsche, Hofmann, Muspratt, Laurent, Gerhardt, and others. Before it had become an article of general use and its manufacture was more of a monopoly than at present, the following table by MM. Laurent and Casthelaz was an evidence of what chemistry had done with coal.†

	Fr.	Cent.	Per Kilogramme = 2 lbs. 3¼ oz.
1. Coal	0	04	
2. Tar	4	10	"
3. Heavy oil	0	20	"
4. Light oil	1	25	"
5. Benzine	2	50	"
6. Rough nitro-benzine .	7	"	"
7. Rectified nitro-benzine	12	"	"
8. Ordinary aniline .	45	"	"
9. Violet & carmine aniline	75	"	"
10. Pure aniline violet in Powder . .	3 to 4000		"

Thus, with coal carried to its tenth power, the price of gold was reached.

The present price of aniline in powder varies from $7.50 to $8.00, and $10.00 per pound. England, France, and Germany supply the American market for the greater part. A commercial list of aniline dyes from the house of Messrs. Toepke & Leidloff, Magdeburg, Germany, is appended:

---

* *Chemical News*, London, 1860. Vol. ii., p. 77.

† *Chemical News*, vol. ix., p. 217. 1864.

Rubin I. [Reddish Red.] } Pure Crystals, dissolving without alcohol in hot water, by boiling five minutes.
Rubin II. [Bluish Red.]

Lilac. [Shade between Rubin and Violet I.]

New Violet I. [Purple of a very Reddish shade.]

New Violet II. [Purple of a somewhat deeper shade of Blue.]

New Violet III. [Purple of a still deeper shade of Blue.]

Parme. [Purple of a very deep shade of Blue.]

Aniline Blue I. [Reddish Blue.]

Aniline Blue II. [Greenish or Night Blue.]

Orange of Aniline.

Green of Aniline, in Crystals.

Red-Brown of Aniline.

Salt of Aniline. For Cotton Goods.

Fuchsine. Reddish and Bluish Red.

PRODUCTS OF THE DISTILLATION OF COALS AT A HEAT OF 700° TO 800° FAH.

The oily products distilled from coals at a high heat, or those produced in coal gas manufactories, have been called tars. However incorrect this appellation may seem to the chemist, it will serve to distinguish those coal tars from the products distilled from bituminous substances at heats just sufficient to expel all the volatile matter they are capable of affording. These are oils. The same description of coals, distilled at the same temperature, and by the same mode, will always yield the same results. The principal products of the decomposition of coals at a gas-producing heat, have been already noticed; but in order to obtain the greatest amount of commercial oils, the heat applied to the distilling vessel should not exceed 800° Fah., while for the production of illuminating gas a temperature of 1000° to 1200° will be required. Nevertheless, it should ever be remembered, that to make the greatest quantity and the purest oils, different coals require different heats, some of them yielding up their oily vapors more readily than others. Therefore, if the same coals which produce the before-mentioned compounds of carbon and hydrogen contained in coal tar, be dry-distilled at a heat not exceeding 750° or 800°, the products will be different in quality and quantity. Instead of benzole, there will be eupion; naphthalin will not be formed, and if formed, the quantity will be small; the quantity of paraffin will be greatly increased, and the amount of creasote or carbolic acid reduced; so that the purification is less expensive. There will be, also, a great change in the quality of the oils. Instead of coal tar naphtha, which cannot be burnt in common lamps without smoke, on account of its being surcharged with carbon, there will be a large amount of oils, with fewer equivalents of carbon, and admirably adapted for illumination, and also denser oils for lubrication. The following are the results of one ton Newcastle cannel coal, distilled for gas and for oils:

DISTILLED FOR GAS. Products.		DISTILLED FOR OILS. Products.	
Coal gas, . .	7·450 cub. ft.	Gas, . . .	1·400 cub. ft.
Coal tar, . .	18½ gals.	Crude oil, . .	68 gals.
Coke, . . .	1,200 lbs.	Coke, . . .	1,280 lbs.

Products of the Coal Tar.		Products of the Crude Oil.	
Benzole, . . .	3 pints.	Eupion, . . .	2 gals.
Coal tar naphtha, .	3 gals.	Lamp oil, . .	22·5 "
Heavy oil, naphthalin, etc.	9 "	Heavy oil and paraffin,	24 "
Total, . .	12⅝ gals.	Total, . .	48·5 gals.

The product set down above as lamp oil consists of several oils combined, which will be noted hereafter.

PRODUCTS OF THE DISTILLATES OF ASPHALTUM, BITUMEN, PETROLEUM, ETC.

The asphaltum of New Brunswick, now called Albert coal, is one of the richest materials ever discovered for the manufacture of oils. Seventy per cent. of the first distillate, after purification, may be brought up to a specific gravity of 0·820, and burned in the ordinary coal-oil lamp. The material contains nitrogen, and therefore yields ammonia. It melts in the retort, and the volatile parts escape at a lower heat than those of coal. This may account in some degree for its greater yield of oils, and their freedom from impurities. From it naphthalin is seldom produced; and although paraffin is found among its products, creasote and other compounds of its class exist but in small quantities, while the illuminating oils are abundant. The oils themselves belong to a series which contains less carbon than ordinary coal oils. They burn freely, and give a clear, white light. The asphalte, or bitumen, of the Dead Sea, affords much oil, mixed with the impurities before noticed. There is present, also, a peculiar volatile oil, which gives even to its purest products an unpleasant smell. This might properly be called *odorine*, although it does not agree with the odorine of Unverdorben.

The bitumen of the Pitch Lake of Trinidad contains sulphur, and sulphuretted hydrogen issues from the pit where the semi-liquid mineral is discharged from the earth. By distillation it also yields a whole series of hydro-carbon oils, some of which have been called naphtha, and represented as $C_6 H_5$; others $C_{20} H_{16}$. It is quite evident that bitumens and their distillates differ materially in their composition, and therefore their value for the manufacture of illuminating oils, or for gas, can only be ascertained by experiment. This bitumen yields 70 gallons of crude oil per ton of 2240 lbs. The impurities in its first distillate are numerous. Among its soluble parts pyroxilic spirit and other products of the distillation of wood have been detected, giving evidence of the vegetable origin of the pitch. All the oils distilled from this substance have a most forbidding smell, which arises from a volatile oil. This oil bids defiance to acids and alkalies, indeed the latter render it more persistent.

The bitumens of Cuba yield from 100 to 140 gallons per ton of the crude oil. These, when purified, are admirably adapted to lamps. A British company shipped the bitumens (chapapote of the Spaniards) to England for the making of lamp and lubricating oils; but the odor followed them, and presented an obstacle of significance. Few of the bitumens of Central and South America have been tested in reference to their composition, or value for hydro-carbon oils. Those of the United States and Canada are beginning to draw the attention of manufacturers in reference to their value in competition with cannel coals and petroleum.

The bitumen of Canada West contains decayed vegetables, and is no doubt the result of petroleum that has long been exposed to the air; 2000 lbs. yielded 109 gallons of crude oils. From this crude product 64 gallons of lamp oils were distilled, and also 18 gallons of heavy oils suitable for lubricating machinery. It differs very essentially from the bitumens of the West India Islands, and the oils require careful purification. The great diversity in the characters of these substances opens an extensive range for chemical research.

Bituminous sands and clays are found at many sites in Central and South America. These, when submitted to dry distillation, afford various quantities of gases and oils, which possess the ordinary characters of bitumen oils. Among those substances may be reckoned the "prairie gas stone" of Illinois, of which glowing descriptions have appeared in newspapers. This is a grey limestone, with pores and cells partially filled with bitumen. By distillation therefore the rock yields hydro-carbon oils, carburetted and bicarburetted hydrogen gases. One sample of the rock gave at the rate of 18 gallons of crude oils, per ton. The bituminous brown coal of Ouachita, Arkansas, has already been noticed.

All these oils, when purified, and when they are of a specific gravity less than 0·850, are extremely fluorescent. When freed from acids they appear yellow by transmitted light, and by reflected light blue. The beautiful hues of the rainbow are sometimes brought out by frequent distil lations and the use of sulphuric acid and caustic alkalies, by which the illuminating oils are frequently injured. It is a peculiar feature of impure coal oils to change color by exposure to the air and light. Oils that come from the worm of the still perfectly colorless will turn yellow, then red, and in a few days a dark brown. Sometimes this change of color begins at the surface of the oil, and pro ceeds downwards until the whole mass is discolored. This arises from the oxidation of the impurities by the atmo sphere. Changes of color also arise from the predominance of an acid or an alkali in the oil, which should be perfectly neutral. The purest oils, when exposed to the direct rays of light, will vary in color, according as the day is bright or cloudy. They possess photographic properties not well understood.

Some of the petroleums, when exposed to the air, evaporate rapidly down to a thick bitumen, others resist evaporation, and "skin over" like linseed oil. Their oils differ from those distilled from coals. They require a greater heat in their distillation, and their vapors are extremely inflammable. These petroleum oils usually commence to boil at 160° Fah., but sometimes at a still lower degree of heat. The lighter or spirituous parts of the charge then begin to distil off, and as the heat is increased the heavier portions come over in succession until the thermometer reaches 565°, when paraffin, if any be present, will begin to appear. It is therefore extremely difficult to obtain any one specified oil, of which the aggregate is compounded. A thermometer fixed in the still indicates the boiling or distilling point of the mass at the time of observation, and nothing more. Each of the oils composing the aggregate collection has a different number of the equivalents of carbon and hydrogen, with which the several boiling points doubtless agree; but the exact rate at which the boiling point does increase, according to the proportions of carbon and hydrogen present in the several oils, has not been accurately discovered.

Laurent has given the composition of some of the oils distilled from bituminous schists as follows:—

Boiling Points.	Carbon.	Hydrogen.
144°	86	14·3
171°	85	14·1
216°	86·2	13·6
304°	85·60	14·5

St. Evrre gives the following:—

Boiling Points.	Carbon.	Hydrogen.
520° and 536°	36	34
485° " 500°	28	26
414° " 428°	24	22
268° " 275°	18	16

Candle Tar.—When the tar resulting from the manufacture of stearine is submitted to distillation, it sends over a series of oils, the chief number of which are good illuminators. Paraffin also appears in the latter part of the operation. Frequently there is the production of much acroleine, the vapor of which produces a burning sensation in the throat and nostrils, and is very unhealthy. These oils are easily purified by alternate washings with sulphuric acid and solutions of the caustic alkalies, with final distillation. They are of a light orange color. The lighter oils are colorless, and by rectification they may be obtained of a specific gravity not exceeding 0·680. The denser oils are superior for lamps.

Caoutchene, or oil of caoutchouc, is produced by the distillation of India rubber, at a moderate heat. A series of light oils, easy of purification, is the result. The vapors are very heavy, and dissolve the resins, shellac, and amber. These oils have been represented as being *caoutchene*, which boils at 72°, Faradayine at 96°, eupione at 124°, and caoutch*ine* at 330°.

Gutta Percha yields oils nearly allied to the above.

CHAPTER VI.

Composition of distilled oils.—Homologous compounds.—Table of the same.—Compounds of Carbon and Hydrogen.—Gaseous compounds.—Homologues obtained from coal tar, coal, bitumen, caoutchouc, etc.

OXYGEN OILS.

BEEORE entering upon a description of the methods employed for the purification of the before-mentioned oils, it is considered necessary to give some account of their component parts and their derivatives. Oxygen enters into the composition of all animal and vegetable oils, unless those oils have been submitted to distillation, which, in general, removes their oxygen and changes their characters. The oils distilled from plants with water are known as essences, or essential oils. They seldom contain oxygen, and are therefore called hydro-carbon oils. The volatile vegetable oils contain oxygen perhaps without an exception.

The oils distilled from the bituminous and oleaginous substances described in the preceding chapters contain no oxygen when they are pure; they are composed of carbon and hydrogen, and are therefore hydro-carbon oils. The greater the quantity of carbon, in proportion to the hydrogen any one of them contains, the greater is its specific gravity, the higher its boiling point, density of vapor, and tendency to smoke when employed for the purpose of illumination. An excess of carbon, however, does no harm to any oil designed for lubrication, but rather gives it consistency and durability. Regarding lamp oils, the greater the amount of carbon they contain the greater will be their illuminating powers, and therefore that is the best lamp, which, when lighted, will decompose the greatest amount of carbon in the flame. It is to the equivalents of carbon and hydrogen contained in oils the attention turns as to a starting-point in this inquiry.

ORGANIC AND HOMOLOGOUS COMPOUNDS.

It is well understood that certain series of organic compounds occur, in which the quantities of carbon, hydrogen, oxygen, and nitrogen increase or decrease, rise or fall, in exact and certain quantities, or number of equivalents. Take, for example, twenty volatile acids, as given by Dr. Gregory, and with a general formula of $C_2\ H_2\ O_4$,* as follows:—

1	Formic acid	$= C_2\ H_2\ O_4$
2	Acetic "	$= C_4\ H_4\ O_4$
3	Propylic acid	$= C_6\ H_6\ O_4$
4	Butyric "	$= C_8\ H_8\ O_4$
5	Valerianic acid	$= C_{10}\ H_{10}\ O_4$
6	Caproic "	$= C_{12}\ H_{12}\ O_4$
7	Œnanthylic "	$= C_{14}\ H_{14}\ O_4$
8	Caprylic "	$= C_{16}\ H_{16}\ O_4$
9	Pelargonic "	$= C_{18}\ H_{18}\ O_4$
10	Capric "	$= C_{20}\ H_{20}\ O_4$
11	Margaritic "	$= C_{22}\ H_{22}\ O_4$
12	Laurostearic"	$= C_{24}\ H_{24}\ O_4$
13	Cocinic "	$= C_{26}\ H_{26}\ O_4$
14	Myristic "	$= C_{28}\ H_{28}\ O_4$
15	Benic "	$= C_{30}\ H_{30}\ O_4$
16	Ethalic "	$= C_{32}\ H_{32}\ O_4$
17	Margonic "	$= C_{34}\ H_{34}\ O_4$
18	Basic "	$= C_{36}\ H_{36}\ O_4$

* *Handbook of Organic Chemistry*, 3d Edition. By William Gregory. London, 1852.

19 Balenic acid	. . . . .	$= C_{38} H_{38} O_4$
20 Behenic "	. . . . .	$= C_{42} H_{42} O_4$
21 Cerotic "	. . . . .	$= C_{54} H_{54} O_4$
22 Melissic "	. . . . .	$= C_{60} H_{60} O_4$

Here we see the quantities increased by the number 2, while the oxygen 4 is constant.

By his able and ingenious researches Laurent discovered a law of substitution, by which one element is replaced by another, according to a perfect and harmonious system. The correctness of this doctrine received confirmation by Dumas, Dr. Hofmann, and Baron Liebig, and its opponents yielded up their views to its facts.

"Of fifteen elements, the equivalents of ten of them, or two-thirds, are represented by whole numbers, that is, they are exact multiples of that of hydrogen, the lightest of them all. They are—

"Hydrogen	. . . .	= 1·0
Oxygen	. . . .	= 8·0
Nitrogen	. . . .	= 14·0
Sulphur	. . . .	= 16·0
Bromine	. . . .	= 80·0
Iodine	. . . .	= 125·0
Fluorine	. . . .	= 19·0
Phosphorus	. . . .	= 32·0
Arsenic	. . . .	= 75·0
Carbon	. . . .	6.0

"If only ten of these were known to us, the law would immediately be assumed that *the equivalents of the metalloidal elements are exact multiples of the equivalent of hydrogen.**

A series of types has therefore been discovered. Those types consist of different elements, and to which other simple substances may be added, or replaced while the ori-

* *Elements of Chemistry.* By M. V. Regnault. Vol. i., p. 347.

ginal type is preserved. The series of volatile oily acids is only one of a number of such series already made out, and to which the oils distilled from oleaginous and bituminous bodies must be added. These series are homologous. Each member of them differs from the others by a certain number of the equivalents of carbon and hydrogen, or by a multiple of them. In their properties these compounds are perfectly analogous, and only differ in degree, and the difference is exactly in proportion to the amount of carbon and hydrogen they contain.

Taking the example given by Dr. Gregory,

"Pyroxilic spirit is	$C_2\ H_4\ O_2$
Alcohol is	$C_4\ H_6\ O_2$
. . . .	$C_6\ H\ \ O_2$
. . . .	$C_8\ H_{10}\ O_2$
Oil of potato is	$C_{10}\ H_{12}\ O_2$

Then the alcohol and pyroxilic spirit differ by $C_2\ H_2$. The oil of potato and pyroxilic spirit differ by 4 $C_2\ H_2$. The compounds between the oil of potato and alcohol have not been discovered.

When a series of substances, especially if derived from the same source, is discovered to have analogous properties, it may be presumed that their compounds are homologous. Although some of the members of the group, or links in the chain, are undiscovered, they may yet be obtained, and the perfect series completed. It is only a few years since two of the acids obtained by the oxidation of alcohol—the formic ($C_2\ H\ O_3$) and acetic $C_4\ H_3\ O_3$)—were known. Now recent discoveries have filled up the series to sixty equivalents of carbon.

The alcohols and ethers, and the acids of their different series, differ by $C_2\ H_2$, or multiples of one or both of these numbers. Still, in all the members of a group there is a family likeness. Here, also, the boiling point, and the density of the vapor, are governed by the proportion of carbon present. Ethyle, methyle, etc., have their derivatives. Each of these derivatives is the starting-point of a series of homologues. M. Dumas, Dr. Gregory, and others, have brought to notice the great analogy between the elementary groups—chlorine, bromine, iodine, potassium, sodium, lithium, etc., and homologous organic groups. Every organic compound belongs to some series in which each individual member of the elementary substances is increased or diminished by certain regular and fixed quantities. The fact may be again repeated, that the oils before described as resulting from the distillation of the different oleaginous and bituminous compounds, are not each a single oil of their kind, but consist of many members, which form a series of oils distinct one from the other. They have the same root, but differ in the branches. Each member of all their several groups contains a different number of the equivalents of carbon and hydrogen, forming chains which rise, step by step, from the solid to the liquid, and from a dense liquid to a light and extremely volatile spirit, and finally to a gas. Again, each of those members is capable of forming entirely new series of compounds, when combined with other elements. As regards the original oily groups, when their components of carbon and nitrogen are the same, their properties will be the same, irrespective of their origin. They will give the same amount of light when burned in lamps, and be equally applicable to useful purposes. This likeness can only be discovered by their specific gravity, boiling points, and, more important than all, by their ultimate analysis by the chemist. As all those oils are capable of affording light, and the term "photogen" applies only to one of them, the appella tion of hydro-carbon, or lamp oils, has been applied to all that are now consumed for illuminating purposes.

As the oils here treated of consist of carbon and hydrogen, some notice may be taken of those two elements. Carbon occurs abundantly in the animal, vegetable, and mineral kingdoms. In its pure and crystallized state it constitutes the diamond. It is the chief substance of plumbago, and frequently forms more than ninety per cent. of anthracite coal. It is essential to the organization of animals, and enters extensively into the composition of minerals, especially the varieties of coal, bitumen, petroleum, etc., and all substances of vegetable origin. Carbon appears also in the gases of coal mines, as carburetted hydrogen, or fire-damp, or carbonic acid, or choke-damp. When organic matter is heated in close vessels, volatile substances are expelled; these consist of carbon, hydrogen, nitrogen, and oxygen; the residue is carbon mixed with the ash—the minerals that enter into the composition of the wood. Carbon is without taste or smell, and insoluble. It resists decomposition, and, when buried in the earth, is imperishable.

Combined with oxygen, carbon forms two gaseous com pounds, carbonic acid and carbonic oxide. Carbonic oxide may be considered a compound radical. It combines with chlorine, oxygen, and the metals. It is a transparent, colorless gas, without taste or smell, and, when inhaled, is fatal to animal life. This gas takes fire, and burns with a fine blue flame, which is often seen on the surface of coals burning in a grate.

Carbonic acid is formed by the respiration of animals, and by vinous fermentation. It is a product of combustion, and is produced artificially by the action of acids upon carbonate of lime. It is a colorless gas, and so much heavier than air, that it may be contained in open vessels. The effervescing properties of wine, beer, soda-water, and some mineral waters, arise from the presence of this acid. It forms the food of growing plants, a part of which they retain in their structures. Another part is expelled, and is found in the atmosphere.

Hydrogen forms one-ninth part, by weight, of water, and exists in vegetable and animal substances. It has neither taste, color, nor smell, and is the lightest substance discovered in nature. It is nearly sixteen times lighter than oxygen, and fourteen and a half times lighter than air. It was, therefore, first employed in floating air balloons. A pressure of a thousand atmospheres has no sensible effect in the condensation of hydrogen gas. Sound moves with three times the velocity in hydrogen that it does in common air, and it refracts light with more power than any other gas. The greater the quantity of hydrogen present in any body, the less will be its weight, or specific gravity. It is thus with the hydro-carbon oils. Hydrogen is also the most inflammable substance in nature; it burns with an almost colorless flame, and great heat. The opinion is entertained by some, that hydrogen is a gaseous metal, as mercury is a liquid metal.

Carbon and Hydrogen, hydro-carbons.—Carbon and hydrogen combine in a great number of proportions, and consequently produce numerous compounds; and as both elements are combustible, their compounds are also combustible and inflammable. By some these compounds are called carbo-hydrogens. At the ordinary temperatures, some of these are solid, such as paraffin, naphthalin, etc.; others are liquid, as the oils of lemons, naphtha, etc. Two of them are gaseous, namely, light carburetted hydrogen gas, and olefiant gas, which are the roots of two, if not more, series of compounds. All these compounds are the products of vegetables, or they are produced from the decay or destructive distillation of organic matter.

Carburetted hydrogen (C, H_2) mixed with atmospheric air is the explosive fire-damp of coal mines, and it frequently issues from the earth through fissures connected with beds of coal, or collections of petroleum. When mixed with twice its volume of oxygen, it explodes with great violence. If mixed with about six times its volume of air, it also explodes. By this mixture gasometers have been blown up with terrible effect.

Bi-carburetted hydrogen, or *olefiant gas* ($C_2 H_2$), mixed with the above and other gases, occurs in coal mines. It is also transparent and colorless. It takes fire readily, and burns with a white flame, giving out much light. It is also the root of an extensive series of hydro-carbons. This gas and the preceding carburetted hydrogen, when pure, form what is known as coal gas, now extensively employed to light cities. Its value depends much upon the quantity of olefiant gas contained in the mixture.

The light produced by the combustion of the hydrocarbon oils is like that of coal gas. It is from gas in both instances. The oils are put in lamps, and inflamed; the gas is produced at the top of the wick, and decomposed instantaneously. In the other instance, the gas is made by heating the coals in retorts, and storing it in gasometers ready for use, and its distribution through pipes and burners. In the benzole, or atmospheric light, the vapor of the hydro-carbon is conveyed in the air to the burner, and there burned as coal gas. The fluctuations in the condensation of this vapor by changes of temperature are impediments to this mode of supplying artificial light.

Homologous series obtained from coal tar. The radical is $C_{10}H_4$; the multiple is C_2H_2.

		Boiling point.	Spec. grav.
$C_{10}H_4$		135°	
$C_{12}H_6$	Benzole	186	850
$C_{14}H_8$	Toluene. . . .	237	870
$C_{16}H_{10}$	Xylole	288	
$C_{18}H_{12}$	Cumole	339	
$C_{20}H_{14}$	Cymole	490 *	

It will be here observed that the boiling point rises 25·5° for every additional equivalent of carbon. By the action of chlorine, bromine, nitric acid, etc., each of the above hydro-carbons forms the root of other distinct and well-defined series.†

Homologous series obtained from the bitumen of Trinidad, distilled at a low heat:

No.	Carbon.	Hydrog.	Sp. grav.	Boiling point.	
1	4	3	0·710	130°	
2	5	4	0·720	155	
3	6	5	0.730	180	
4	7	6	0.740	205	
5	8	7	0.750	230	Embracing the hydro-carbon oils suitable for lamps. Specific gravity of the whole when mixed 0·819.
6	9	8	0.760	255	
7	10	9	0.770	280	
8	11	10	0.780	305	
9	12	11	0.790	330	
10	13	12	0.800	355	
11	14	13	0.810	380	
12	15	14	0.820	405	
13	16	15	0·830	430	
14	17	16	0·840	455	
15	18	17	0·850	480	
16	19	18	0·860	505	

* Generally represented as $C_{20}H_{16}$.

† See Gregory's *Handbook of Organic Chemistry*, 3d edit., p. 129.

No.	Carbon.	Hydrog.	Sp grav.	Boiling point.
17	20	19	0·870	530
18	21	20	0·880	555
19	22	21 Paraffin	0·890	580*

A sample of petroleum from Western Virginia produced a series of oils agreeing with the foregoing, but there is much diversity in the character of the petroleum in regard to their densities and boiling points, and it is remarkable that the denser oils require a higher degree of heat for their distillation than oils of the same specific gravity obtained from coals.

The bitumen of Cuba, Albert coal, bituminous shale of Albert county, the petroleum of Virginia, and candle tar, produce the same series of hydro-carbons.

The series obtained from Breckenridge coal, distilled at an average heat of 780°, was as follows:

No.	Carbon.	Hydrog.	
1	4	2	Supposed to exist, but not condensed.
2	6	4	
3	8	6	Embracing the hydro-carbon oils suitable for lamps when mixed. Spec. grav. 0·819. (Nos. 3–9)
4	10	8	
5	12	10	
6	14	12	
7	16	14	
8	18	16	
9	20	18	
10	22	20	Paraffin.

A coal from Kanawha, Virginia, when distilled at a heat of 900°, gave part of a series thus:

* There is some diversity of opinion regarding the condition of a liquid when it is said to boil. In taking the boiling points of the foregoing, the oils were allowed to be in a state of full ebullition, and distillation commenced at the times when the heat was recorded by the Thermometer, the barometer being at 30—Fractions were omitted.

No.	Carbon.	Hydrog.
1	8	4
2	12	8
4	16	12
4	18	16

Caoutchouc was distilled at a moderate heat, and the following was the series produced:

No.	Carbon.	Hydrog.	Sp. grav.	Boiling point.
1	8	7	678	94
2	9	8		
3	10	9		
4	11	10		
5	13	11		
6	14	12		

Other series of hydro-carbons might be laid down; but the foregoing are sufficient to demonstrate the existence of a system which cannot be carried forward to perfection without great labor and research. This system is being gradually extended to every branch of chemistry, and is bringing the science into a beautiful harmony with mathematics, and its kindred study, Astronomy.

To the manufacturer it is of the first importance. It teaches him that he has to deal with a great variety of compounds. An increase in the degree of heat employed in his operations will change the properties of the products, increase the proportion of carbon, and defeat him in his objects. A temperature too low will give results to disappoint him. He cannot fail to observe the different proofs at which his oils flow from the still, and the constant increase of heat required to produce them in the process of refining and purifying; and having obtained even an indistinct view of the point he would reach, his skill and experience will bring to him that knowledge of his art he desires.

CHAPTER VIII.

Oxidation of the impurities contained in crude hydro-carbon oils.—Action of acids, alkalies, and other agents.—Sulphuric acid, nitric acid, permanganate of potash.—Methods of purification.—Extracts from patents, etc.

WHEN oils were first distilled from coals, few attempts were made to free them from their offensive odors, or remove their coloring matters. The only mode practised was fractional distillation, which is altogether quite ineffectual for that purpose. Although the oil made by the Earl of Dundonald in 1781 was burned in lamps, it does not appear that any process of purification was practised at that time. The earliest mode of purifying petroleum was simply to distil it with water, and this is more beneficial than some of the modes practised in the present day, by which the characters of the oils are changed and their illuminating powers deteriorated.

The great number of impurities contained in the oils distilled from coals, whether from coal tar or crude coal oil, renders their purification somewhat difficult, expensive, and uncertain. The varieties of coals and other substances employed to obtain hydro-carbon oils, the fluctuations of heat in distillation, and varying qualities of reagents, will ever require the care and skill of the practical chemist to overcome them. Much has been done in the purification of those oils, much is still to be performed before they are made perfect, namely, free from all offensive odor, and free from color. The great difference observed in the qualities of the oils in the market arises less from the different modes by which those oils are treated, than from the properties of the coal from which they were distilled.

ACIDS, ALKALIES, AND OTHER OXIDATING AGENTS.

Acids, alkalies, peroxide of manganese, permanganate of potash, bichromate of potash, etc., have been unsparingly used in the purification of hydro-carbon oils, on account of their oxidating properties. The object of chemists has been to impart oxygen to the impurities, by which they separate themselves from the oils, and generally fall to the bottom of the vessel that contains them.

The oxidation of organic compounds takes place in several ways. In combustion atmospheric oxygen is aided by a high temperature. If the supply of air be deficient, as in the case of a burning lamp, the hydrogen, from a greater attraction for oxygen, is oxidated, and the carbon of the oil appears in smoke or soot. The decay of wood is produced by oxidation, and ulmine is the result. So also in some of the impurities in hydro-carbon oils; their combination with oxygen gives them new characters, by which they no longer remain with their native liquids. Reagents may be applied to oils that will not separate from them until exposed to the heat of distillation. By its oxidating properties permanganate of potash converts sugar into oxalic acid. Bichromate of potash diluted with sulphuric acid converts salicine into the hydruret of salicile, or oil of spirea. Organic substances are oxidated by the atmosphere, and its action promoted by a high temperature. Hot air has therefore been forced through hydro-carbon oil during the process of purification, and, in some instances, with advantage.

Action of sulphuric acid.—In general, when sulphuric acid is applied to organic compounds (and such are the oils under consideration), it decomposes, or chars them. By the aid of heat its effects are more powerful, and it transmutes starch and lignine into grape sugar. Its action upon naphthalin and other compounds of carbon and hydrogen has been before noticed. Paraffin is not sensibly affected, when boiled with sulphuric acid. For this reason it is employed in the purification of that substance, as it absolutely burns out all its impurities. Sulphuric acid, or oil of vitriol, is now universally used in the purification of coal oils, by which some of their impurities are converted into tar, or rendered soluble in water. The acid may be separated from the tar by distillation. This acid always decomposes a part of the oils in proportion to its strength and the quantity employed. It is a powerful purifier. It removes one kind of odor and substitutes another less disagreeable. How far it changes the characters of the oils has not been determined; but in some instances, when it is used in large quantities, there can be no doubt it produces what may be called *sulpho-oils*, which are unchangeable by the use of alkalies. Certain it is that these sulpho-oils are quite dissimilar to the natural oils obtained by the fractional distillation of coal oils, and are inferior to them for the purposes of illumination. The powerful effects of the before-mentioned acid in removing impurities from the distillates of coal, and its cheapness, have brought it into general use.*

Action of nitric acid.—The operations of nitric acid upon organic substances are very numerous. It usually, if not always, produces one or more acids. From gum there

* The average specific gravity of commercial sulphuric acid is 1·800. It sometimes contains nitric acid.

comes mucic acid; from indigo, indigotic and nitro-picric acids; from stearic acid, margaric acid, etc. Laurent has clearly described the action of nitric acid upon naphthalin.

Benzole admits of having its hydrogen replaced by one, two, or three equivalents of nitric acid. This remark applies equally to eupion and all the lighter products distilled from coals, petroleum, etc. All these compounds have an aromatic odor. As an instance, when benzole is saturated with fuming nitric acid, and water is added to the hot solution, nitro-benzole subsides as a yellow oil with the odor of cinnamon. It is sold as the oil of bitter almonds. Other light hydro-carbons give similar results, and a great number of oils, useful for perfumery and cookery, may be produced from them.

As an oxidator nitric acid is more powerful than sulphuric acid; but it exerts a greater action on the oils themselves, changing them into *nitro* oils, and removing them further away from the natural products of the material first employed.

Permanganate of potash must be included among the materials used for oxidating the impurities contained in distilled oils. Its effects are feeble when compared with those of sulphuric acid, and its price is too great a drawback on the profits of the manufacturer.

METHODS EMPLOYED FOR THE PURIFICATION OF HYDRO-CARBON OILS.

The earliest writers on the production of oils from coals and other analogous substances, did not describe any very satisfactory mode by which those oils could be purified. Selligue was perhaps the first to supply a method for this purpose; and it appears in the voluminous specification of his patent.* He commenced by agitating the oils with sulphuric, muriatic, or nitric acid. The agitation was continued for some time, so that every particle of the oil should be brought in contact with the acid, and a certain change of color had taken place. His agitators were of peculiar construction, and he has described them at length. After the oil and acid had been allowed time to separate, the former was decanted and washed with soap-maker's lye, proof 36° to 38° Baumè. Thus a part of the coloring matter was precipitated, although some of the lye was subsequently permitted to go into the still with the oils. Fractional distillation was also resorted to, which with variations in the above mode enabled the chemist to produce oils of good quality. The specification of Selligue was written with great care; but his operations were complex and expensive. The alternate use of acids and alkalies forms the principal feature in the purification of those oils at the present time.

MANSFIELD'S PROCESS.

In 1847 C. B. Mansfield of Cambridge, England, obtained a patent for the "purification of spirituous substances and oils" derived from coal tar, &c. Of the products of coal tar he describes five, namely, alliole, benzole, toluole, camphole, mortuole, and nitro-benzole; for each of these classes he modified the treatment. To alliole and benzole, he applied diluted sulphuric or hydrochloric acid, and agi tated the mixture, which was allowed to settle, when the acid and impurities were drawn off. The spirits and oils

* Specification No. 10,726, English Patent Office. Translated by Du Buisson.

were then agitated with water, which was also afterwards removed, and the spirits and oils placed in a vessel of fresh burnt lime, and finally rectified by distillation. The toluole, etc., were purified by a similar method, except that stronger and greater quantities of the acids were employed, and the number of distillations increased. The specification of this patent is also of great length, and directed to objects foreign to the purification of the oils derived from bituminous substances.

YOUNG'S PROCESS.

This alleged improvement consists in treating bituminous coals in such a manner as to obtain therefrom an oil containing paraffin, which is denominated "paraffin" oil, and from which Mr. Young obtains paraffin. He employs "Parrot coal," "cannel coal," and "gas coal." These are broken up to about the size of a hen's egg, and distilled in common gas retorts with worm pipes and the ordinary refrigerators of stills, the water in them being kept at a temperature of about 55° Fah., by a stream of cold water entering the worm cistern. The retort is kept at a *low red heat.* The retort is heated up gradually, and the product is an oil containing paraffin.

The crude oil is put into a cistern, and steam heat applied up to about 156°. This separates some of the impurities, and the oil is run off into another vessel, leaving the impurities behind. The oil is then distilled in an iron still with a worm pipe and refrigerator, the water in the latter being kept at 55° Fah. The oil thus distilled is then agitated with ten per cent. of oil of vitriol one hour. It is then allowed to settle twelve hours, when it is drawn off from the acid and impurities into an iron vessel, where it is again agitated with four per cent. of the solution of caustic soda of specific gravity 1·300. Six hours are again allowed for the alkali and impurities to settle, when the oil is again drawn off and distilled with half its bulk of water; water being run into the still from time to time to supply the quantity distilled off. The lighter oil comes over with the steam, and is employed for illumination. The oil left in the still is carefully separated from all water and put into a leaden vessel, and there agitated with two per cent. of oil of vitriol. It is then allowed to settle twenty-four hours. This oil is then run into another vessel, and to every 100 gallons there are added twenty-eight pounds of chalk, ground up with water into a paste. The oil and chalk are agitated together until the oil is freed from acid. After it nas remained a week at rest, it is used for lubricating machinery, and may be mixed with animal or vegetable oils for that purpose. To obtain the paraffin the oil containing it is brought down to a temperature of 30° Fah., when paraffin will crystallize and separate itself from the oil, or it may be filtered and finally submitted to pressure. Again it is agitated with its bulk of oil of vitriol, and the operation repeated until the acid ceases to be colored by the paraffin, which is kept melted during the operation.*

KEROSENE PROCESS.

The specification describes the process for obtaining oils, denominated Kerosene, from "bitumen wherever found." The Kerosene consists of three distinct hydro-carbons,

* Extracted from the patent of James Young.

namely, A Kerosene, B Kerosene, and C Kerosene. The C Kerosene, or that which is employed in lamps, may be formed by an admixture of the light with the heavier oils, until the specific gravity is raised up to about 0·800, water being 1000. The first part of the process consists in submitting the raw material to dry, or decomposing distillation, in large cast iron retorts at a temperature not exceeding 800°. The condensation of the vapors is effected in iron pipes, or chambers, surrounded by water.

"The liquid products of this distillation are heavy tar and water, or ammoniacal liquor, which lie at the bottom of the receiver, and a lighter fluid which floats above them." The heavy fluids and the light are separated by drawing off one from the other. "The heavy liquids may be utilised or disposed of advantageously; but they have no further connexion with this process." The light liquid is submitted to re-distillation at the lowest possible heat, in a common still and a condenser. The products of this distillation are a light, volatile liquid, which accumulates in the receiver, and a heavy residuum left in the still, and which may be added to the heavy liquid impurities of the first distillation.

The light liquid is transferred from the receiver to a suitable vessel or vat, and mixed thoroughly with from five to ten per cent. of strong sulphuric, nitric, or muriatic acid, according to the quantity of tar present. Seven per cent. is about the average quantity of acid required. The preference is given to sulphuric acid. With the acid and oil, from one to three per cent. of the peroxide of manganese is added, and the whole thoroughly agitated together. The mixture is allowed to stand undisturbed from twelve to twenty-four hours, in order that the impurities may subside. The light, supernatant fluid is now drawn off into another vessel. The distillate is then mixed with two per cent. or more of freshly calcined lime, which takes up any water that may be present, and neutralizes the acid. The oil is then distilled, and finally rectified, if necessary. The product is kerosene, the lightest part of which is called A kerosene, and the two succeeding parts B and C kerosene.*

The above mode has been much improved by the use of steam, introduced into or above the oils during their distillation, by diminishing the quantity of acid and washing with water. The latter removes much of the soluble impurities. The A kerosene is perfectly colorless, and has a close analogy to eupion. The remaining hydro-carbon oils are of a light straw-color. They burn freely in lamps, without incrustation of the wick.

There are a number of oil manufactories in Germany. In some of these shales are used, in others cannel coal. The coal is usually broken into small pieces, and when it contains sulphur it is moistened with lime-water. The coal is then thoroughly dried in a furnace constructed for the purpose. The dried coals are distilled in common gas retorts, the eduction pipes of which open at the ends opposite their heads. In some instances the flame of the furnace is not permitted to strike the sides or upper surface of the retort.

Paul Wagenmann, of Bonn, Rhenish Prussia, in his patent, states as follows:

"My improvements consist in breaking the coal or bituminous slate in pieces of about the size of a walnut; and if they are very sulphurous I sprinkle them with lime-water. They are then taken to a drying-furnace of the

---

* Extracted from copies of the kerosene patents.

following construction: A space by preference of two hundred feet in length, and twenty feet in width, is intersected by walls of two feet high; at the distance of every four feet these walls are bound together by arches of one brick thick, and on these arches the coals and bituminous slate are spread.

"The space below the arches is filled up with the residue from the retorts.

"The coals or bituminous slate, when dried, are distilled in retorts which are so far different from those used at the gas-works, that the pipes for letting out the produce of distillation are on the opposite ends to those where the doors are. Over each fire are two retorts, each by preference, of about eight feet long, and two feet wide, with an opening of five inches, to let out the produce of distillation. The fire runs below the retorts in a direction from front to back, the fire-bars only extending part of the way. I prefer to arrange a stack consisting of eight fires and sixteen retorts around one chimney, by which means I am enabled to lead the flame from one fire to the others, and by that means to heat the retorts by a graduated heat.

"The products of distillation of the sixteen retorts meet together in one iron pipe about eighty feet long, and two feet diameter, which is surrounded by another, so that cold water can run between the two pipes for cooling. The gases, after having passed this pipe, enter into cylinders, about twelve feet in height and four feet in diameter. These cylinders are filled with iron wire chips. The gases, after having passed the cylinders, pass through another iron pipe, forty feet high into the air, which pipe, to regulate the draught, is furnished with a regulator.

"It is important that the produce of distillation should not be conducted so as to produce pressure in the retorts.

"The produce of distillation runs into a general reservoir, and the reservoir is so arranged that the condensed productions will have an average heat of 30° centigrade. The oils separate themselves here from the ammoniacal water. The ammoniacal water is thrown over the cooled residue of the drying furnaces, and mixed with it, which produces a very good manure. The tar, after being separated from ammonia, is distilled, and the product of distillation is cooled by the means of a lead pipe, standing in a cooling apparatus, the water for cooling being kept always lukewarm. The product of distillation is divided into three qualities: No. I. from the beginning of the distillation to 0·865 specific gravity. No. II. from 0·865–0·900 specific gravity. No. III. from 0·900–1·930 specific gravity.

"The produce No. I. is mixed with sulphuric acid and hydrochloric acid, at a temperature of 25° centigrade. Three hours afterwards the oil is taken off and washed with a solution of caustic soda, at 60° centigrade: it is left two hours and then separated from the solution and distilled. In the still is mixed a concentrated solution of soda. After the distillation the oils are light yellow, and give an average weight of 0·815–0·825 specific gravity. To correct the smell I wash the oils again with sulphuric acid and hydrochloric acid, separate them from the solution, and wash with concentrated solution of soda.

"The oil No. II. is treated the same as No. I., but with different quantities of acids, and at a temperature of 35° centigrade. The product, after the distillation, is a lighter oil.

"The oil No. III. is the product for the preparation of the finest oil and paraffin candles. The oil is treated with sulphuric acid and hydrochloric acid, at a temperature of

83° centigrade, and allowed to stand; it is then separated from the acids, and washed with a solution of soda, at a temperature of 60° centigrade, and distilled. The oil contains paraffin, and is taken to a cool cellar at an average temperature of 12° centigrade, where it remains in iron butts for eight days. After this time the paraffin is separated from the oil by means of a centrifugal machine and cast in cakes, and pressed in a cold hydraulic press; afterwards melted and mixed with sulphuric acid, then separated and washed in water; it is then heated and cast in cakes, and again pressed by a heated press; afterwards again melted and mixed with sulphuric acid at a temperature of 70° centigrade; the acid is drawn off, and the paraffin is washed in water, after this it is melted with stearine."

In some instances the retorts are placed in a circle around the chimney, and two of them are heated by one furnace. The gaseous products of the distillation are conducted into a large iron pipe, upon which a stream of cold water plays constantly to produce the necessary condensation. The uncondensed gases escape at the end of the condensing pipe and are lost. The oils and other liquid products of the distillation flow into a cistern, whence they are pumped for purification.

Having been separated from the aqueous products the oils are submitted to the purifying process. Some chemists have the oils mixed with four per cent. of sulphate of iron, in cast iron cisterns, supplied with agitators worked by machinery. Next the charge of oil is distilled, and for this purpose various expedients have been resorted to. Some distil *in vacuo*, others employ common, or superheated steam. The latter obtains the preference, especially for the heavy oils.

The distillate is usually divided into two parts. The first is permitted to run from the still until the specific gravity comes up to 0·870. The second part embraces all the remainder of the distillate. The first part is then agitated for hours with six per cent. of concentrated sulphuric acid, one-eighth per cent. of bichromate of potash, and one-half per cent. of hydrochloric acid. The second part is treated in the same manner, except that the sulphuric acid is increased to eight per cent., with one-sixth per cent. of bichromate of potash, and one per cent. of hydrochloric acid. After the acid impurities, etc., have subsided they are drawn off and the oils are agitated two hours with lye and steam.

The oils are then distilled, great care being taken that they should not "boil over." By this mode lamp oils, heavy oils, and paraffin are produced. The paraffin is put in a cool place and allowed to crystallize in the usual manner.

At Bitterfield the coal is broken into small pieces and distilled in elliptical retorts eight feet in length. The discharge pipe is of large size and opposite the head of each retort. Pressure upon the material while it is undergoing distillation is avoided as much as possible. The purification consists in the alternate use of acids and solutions of caustic alkalies.

Dr. Vohl, of Bonn, commences the distillation of paper coal at a low heat, which is gradually raised up to a red heat, and he remarks that slates containing twenty-five per cent. of water yielded the largest amount of oil. The author has observed the same fact in the distillation of bituminous shales imported to New York from Pictou, Nova Scotia. When the retorts are first charged with those shales, steam is generated from the water contained in them.

With the steam some of the lighter oils are distilled over, and with it condense. The effect is quite similar to that produced by admitting steam into the retort at the commencement of the decomposing distillation. In both instances the quantity of oils is increased.

Brooman's Patent.—Among the list of patents for the purification of hydro-carbon oils, this patent, which is dated London, February 28, 1856, has been overlooked, with several others of equal importance. The patent is for "improvements in treating bituminous shale, Boghead mineral, and other like schistose bodies, in order to obtain various commercial products therefrom."

The schistose bodies are first decomposed in common retorts. The receiver is placed at some distance from the retorts, and receives through pipes a part of the gas generated in them. Condensation is effected in refrigerating pipes kept cool by water. The oils are treated in agitators, or purifiers with sulphuric acid and caustic soda, and then distilled over again. The light oils are separated from the heavy for illuminating purposes by distilling them down to proof 32° (Gay Lussac's Areometer), all that remains is separated from the paraffin. For this purpose the heavy oil is placed in refrigerators with double bottoms and exposed to a low temperature, by which the paraffin is separated. The remainder is gathered into bags and subjected to pressure, to remove whatever oil it may contain.* The products represented as being obtained by this mode are—

"1. Essential oil.

"2. An oil for lighting purposes.

"3. A fatty, unctuous oil, for lubricating machinery.

---

* Journal of Gas Lighting (London), Sept. 16, 1856.

"4. A liquid tar, for lubricating purposes.

"5. A solid tar.

"6. A 'black,' which may be used in the manufacture of printers' ink.

"7. A 'black,' having the properties of animal black.

"8. Paraffin.

"9. Ammoniacal water, containing six per cent. of liquid ammonia."

There is some obscurity in the specification of this patent; still the practical manufacturer will readily understand, from the above, the nature of the process employed.

Bodmer's patent is dated London, February 4th, 1856.*

"Tars are taken which have been produced by the distillation of coal at a *high* temperature, such as are made in the manufacture of coal gas. This tar, being the cheapest at present, is therefore preferred; but tars produced in a similar manner, at a high temperature, from shale, peat, wood, and from bones, or other animal substances, will answer the purpose. These tars are placed in an ordinary still, into which the bulb of a thermometer is placed, and connected with a worm immersed in water: this water is kept regularly at a temperature of between 60° and 80° Fah., throughout all the distillations. The heat of the still is raised by fire; and when the thermometer in it rises to 300° Fah., the instrument is removed, and the products of distillation above 200° are run into another vessel, and kept separate from the products of distillation below 300°. The latter are rejected as unfit for the purpose. The tar is distilled to dryness, which is known to have taken place when products cease to run from the condenser, the heat being always kept up." "The oil obtained from the coal

---

* *Journal of Gas-Lighting* (London).

tar is purified as follows: This oil is put into a leaden tank, and to each five hundred gallons is added ten gallons of commercial brown sulphuric acid, of strength 140° Twaddle, or about 700 specific gravity, and they are well agitated together for one hour. The vitriol is allowed to subside, which will take place in ten or twelve hours, and is drawn off by a stop-cock placed at the bottom of the tank. Another ten gallons of brown sulphuric acid is then added to each five hundred gallons of the oil, and agitated for four hours. The oil, after subsidence, is removed to an iron vessel, and to each hundred gallons is added ten gallons of a solution of caustic soda, marking 70° Twaddle, or weighing 13½ lbs to the gallon. These are agitated together thoroughly for ten or twelve hours; and it is preferred to keep the temperature of the oil in this tank up to 80° Fah., both during the agitation with the caustic soda and afterwards, for ten or twelve hours. The clear oil is then removed into a still, and to each hundred gallons is added about twenty lbs. of the soda ash of commerce, 20 lbs. of slacked lime, and four gallons of water, or 40 lbs. of caustic solution of soda, marking 70° Twaddle—or by measure, three gallons, weighing 13½ lbs. to the gallon, are taken for each hundred gallons of oil put into the still, and heat is applied. In general, no oil will come over until the heat of the still has reached 300° Fah.; but if any should come below this temperature, it is rejected. When about eighty per cent. of the oil put into the still has been obtained, the process is stopped. The product of distillation is the improved lubricating oil, which is named 'new tar oil.' It may be used either by itself or mixed with other oils, fats, greases, and soaps."

P. G. Barry places the oils in wooden tanks lined with lead. In these tanks the oils are agitated with five per cent. of their weight of sulphuric acid, during a period of three hours. After the acid and impurities have settled, the oils are drawn off into a second purifying vessel, and there agitated with five per cent. of their weight of caustic alkali, or with lime water sufficient to remove all the acid present in them. After the alkaline mixture has subsided the oils are again distilled.

Bancroft obtained an English patent for the distillation of hydro-carbon oils from the petroleum of Burmah. He admits high pressure steam at fifty lbs. to the square inch into his stills, and places a fire beneath them until all the eupion is distilled over. This part of the distillate being removed, the fire beneath the still is increased, and the steam forced on, until about ninety per cent. of the charge is distilled off. At the close of the operation much paraffin appears, which renders it necessary that the condensing pipes should be kept at a temperature not less than 90° Fah. In several instances the cooling down of the condensing apparatus has led to the bursting of the still.

A process is recorded in *Le Genie Industriel*, and represented as being the invention of Messrs. Dumoulin & Cotelle, by which the heavy coal oils are made to burn in lamps without smoke or odor. In a close vessel they place one hundred lbs. of *crude* coal oil, twenty-five quarts of water, one lb. of the chloride of lime, and one half lb. oxide of manganese. The mixture is thoroughly agitated. After a repose of twenty-four hours the clear oil is decanted and distilled. Next the one hundred lbs. of coal oil are mixed with twenty-five lbs. of rosin oil, and this is considered the best part of their mode. This last mixture may be distilled if necessary. From the high per centage of carbon in the heavy coal oil, and also in the rosin oil, it will appear theoretically that this mixture cannot burn without smoking in any of the ordinary coal oil lamps, and this is found to be the fact in practice. In an argand lamp with a short-topped wick, and a button over the inner air tube, or in the camphene lamp, the above oil will burn with a short flame and brilliant light, and so also will the rosin oil, or the heavy oil, mixed or unmixed; but those lamps are rapidly falling into disuse, being supplanted by the kerosene, or coal oil lamp.

It has been already stated that upwards of one hundred patents have been granted for alleged new methods of manufacturing and rectifying oils distilled from coals and other bituminous mineral substances; and upwards of forty patents have been issued for retorts and other apparatus connected with this branch of industry. A description of the various methods and similarities of operation, with the extraordinary and unphilosophical fancies set forth in some of those patents, would not interest the practical man nor the general reader. The extracts drawn from the foregoing patents have therefore been deemed sufficient for this brief technological treatise, and to direct the manufacturer of oils to the valuable discoveries now placed at his hand.

The preceding part of this chapter will have shown that upon a few leading, and, as it is supposed, essential operations, all the patentees appear to agree. Upon non-essentials they differ as widely as persons do in matters of far higher importance.

It is conceded at the present time—

1st, That the crude coal, or other material, must first be submitted to dry, or decomposing distillation, and that a moderate degree of heat will produce more and better oils than a high temperature.

2d, That the use of a strong acid is necessary in the purification of such oils.

3d, That the acid must be succeeded by the use of an alkali.

4th, That it is necessary to distil the oils after the use of the acid and alkali.

It will be perceived by the foregoing extracts, from patents for the manufacture and purification of oils, that distillation, acids, and alkalies, form the basis of every alleged invention; but upon the quantities, the modes of application, and the minor details of working, there is much disagreement; and persons unskilled in chemical science have frequently introduced some peculiar mode in the application of those agents, to give novelty to their patents, or to satisfy their employers of their superior skill.

The oils from different coals require different treatment. The oils of Albert coal (asphaltum), Boghead and Breckenridge coal are easily purified; while the oils from the ordinary American, English, and Scotch cannels, require more skill, and cost more to bring them up to a fair standard among the hydro-carbons sold in the market.

The author has made more than two thousand experiments in reference to the manufacture and purification of oils distilled from coal, petroleum, and other materials. From long practice, and the improvements introduced by others, he ventures to lay the following plan before his readers, as being generally applicable to the distilled oils of coal and bitumen. Petroleum will be noticed in the sequel. Regarding the purification of those oils, the present is the age of experiment. Improvements are constantly advancing, and some time may elapse before their manufacture is brought to perfection, and the distilled hydro-carbon oils attain that commercial and economic value they are destined to reach.

CHAPTER VIII.

Buildings and Machinery.—Method of Manufacturing and Purifying the Oils distilled from Coals and other Bituminous Substances, and the Products derived therefrom.—Distilling by Steam.—Continual Distillation.—Paraffin.—Lubricating Oils.—Purification of Petroleum.—Petroleum Refinery.—Estimate of Cost.—Hydrometer and Pyrometer.—Cements, etc.

DISTILLERY FOR COAL OILS.

BEFORE any suggestions are made in reference to a proper mode of manufacturing and purifying the hydro-carbon oils, the construction and arrangement of the manufactory itself require some notice. It is very desirable, in all cases, that the buildings constituting the establishment should be constructed of stone and brick, with iron roofs. The occupation of wooden buildings is unsafe; when they are employed, great care is necessary. Every preservative against fire, by the use of non-combustible material and the command of water, should be planned for at the onset of construction.

When coal is to be distilled in retorts, the retort house should be separated from the distillery, or refining house, and all crude materials and marketable oils should be kept in separate stores, away from the operating part of the establishment. Receivers of the products of the retorts are advantageously situated underground. A steam pump, communicating with cisterns of water, and supplied with hose capable of reaching every building, should be always ready for action, while at the same time it performs the offices required by the manufactory.

Between the stills and the several worm-tanks and receivers it is necessary to erect a strong brick or stone wall, through which the connecting pieces between the stills and the worm pass. It is also desirable to separate the stills one from the other by partition walls. During the distillation, and especially at its commencement, a light hydro-carbon vapor frequently escapes at the lower extremity of the condensing-pipe. This vapor is highly inflammable, as well as the lighter oils that accompany it. No fire should, therefore, ever be permitted in the body of the refinery.

The agitators should be placed at convenient heights to permit the oils to flow from the acid cisterns into the tanks where they are to be washed with the alkali, and to run thence into the stills. A good arrangement of the machinery is of much consequence; and, above all, the most rigid cleanliness should be observed in every operation connected with the manufactory. An abundant supply of clean, fresh water is absolutely necessary.

The plan and sections (pp. 148–149) represent approved arrangements of the building and apparatus of a coal oil distillery. The number of distillations being much greater than those for petroleum refining, it requires more stills to do the work. The agitators, L, are in number three, being one extra to permit of repairs to the others. The receivers, M, are vessels in which the crude oil pumped from the retort vat is settled. The fourth still, E, may be used as a rectifying still for the eupion. The transferring of the oil from vessel to vessel is effected by the pumps, J, when their levels do not admit of drainage. The dimensions of Petroleum refinery, R, contained in the estimate of cost will serve as a guide to the builder. In the stills there should be room allowed for the ebullition and foaming;

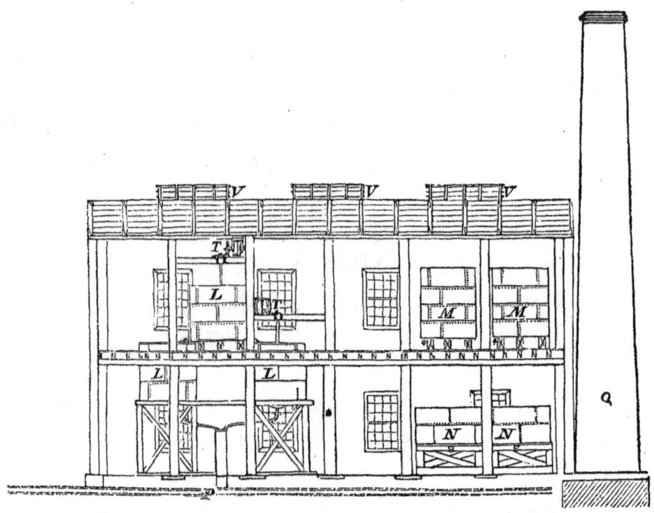

Section on Line C—D of Plan.

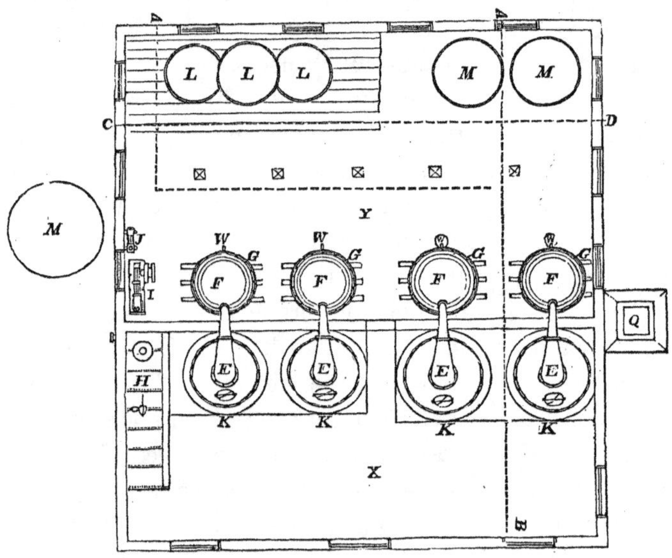

Coal Oil Refinery Plan.—600 gals. Capacity per Diem.

six, eight, or ten inches below the top of the kettle, or that part of the still below the dome, is the proper line for working contents. The agitators should contain the proposed charge of oil, and fifteen per cent. over. The cost of the work may be estimated from that of the Petroleum Refinery.

Section of Broken Line A—B of Plan.

REFERENCES.

E. Stills.
F. Worms.
G. Worm tanks.
H. Boiler.
I. Engine.
J. Steam pump.
K. Still furnace.
L. Washers, or agitators.
M. Receivers.
N. Market tank.
O. Syphon of still pipe.
P. Drain.
Q. Chimney.
R. Water pipe.
S. Steam pipe.
T. Washer gearing.
U. Pipe from agitators to stills.
V. Ventilators.
W. Tail pipes.
X. Still house.
Y. Refinery.

Coals.—The crude oils distilled from coals differ greatly in yield and in quality. It will be observed in the table given in a preceding chapter, that a few varieties will produce over a hundred gallons per ton. Some cannels will not yield over fifty, and others thirty gallons per ton. The qualities of the crude oils also differ. Some afford large quantities of paraffin, or heavy oil, and but a small percentage of lamp oil. Others yield much eupion. In the purchase of coal lands, or coal for the manufacture of hydro-carbon oils, an accurate test of the coal is necessary. Coal oil works have been erected at coal mines, where the coal itself is almost worthless for oil-making.

Bitumens.—The preceding remarks are also applicable to asphaltums and bitumens. The tars of candle manufactories also give different results, and yield some heavier and some lighter oils.

Retorts.—Different retorts also produce different results. When the discharge-pipe is high, there will be less crude oil; but the oil will be lighter and purer. Pressure upon the charge and its vapors during distillation will diminish the yield; and where the condensation is imperfect, a part of the lighter oils will escape with the gas. The revolving retort has the advantage of distilling coals and shales in less than half the time required for stationary retorts. The yield is also large; but the crude oil is impure from the quantity of dust produced by the agitation of the material during its dry distillation. More important than all is the amount of heat applied, which should, as an ordinary rule, not exceed 800° Fah. Before the coal, asphaltum, or any bitumen is thrown into the retort, it is advantageous to break it into small pieces. Large masses seldom discharge all the volatile matter from their central parts.

Condensers.—By removing the heat that attends the vapors and gases produced by distillation, their particles are brought into closer proximity, and all pass from the gaseous into the liquid or solid state, except the permanent gas, which is incapable of condensation by ordinary means. Condensers are usually metallic worms immersed in water, which is kept at the desired temperature by the admission of cold water. It is quite immaterial whether the condenser is a long metallic pipe, a series of pipes, or an open chamber, if it be of sufficient dimensions, and kept at a low temperature.

Receivers—at coal oil manufactories—are tanks, usually sunk in the earth, to allow a descent of the oils from the condensers. They should be closely covered, to prevent the evaporation of the lighter oils, which, in warm weather, is very rapid. The gas which remains uncondensed should be conveyed to a gasometer, and there stored for fuel, or it may be purified in the manner of ordinary coal gas, and employed for lighting. In general the gas is allowed to escape, especially where fuel for the manufactory is cheap. It is admirably adapted to the distillation of oils, and, with proper burners, a high degree of heat may be obtained.

Precipitation or settling.—With the crude oils that flow into the receiving tank there is always a quantity of water, or ammoniacal liquid combined with some carbonaceous matter and other impurities. When the coal, or other material distilled in the retorts, is very moist, the water in the receiver will sometimes amount to twenty per cent. of the distillate. To remove those impurities it is most advantageous to pump the whole into a second receiver, or tank, which should be elevated a little above the working level of the stills it is designed for.

The crude oils and their impurities should next be heated up, by means of a steam coil placed in the bottom of the tank, to 90° or 100° Fah. The ammoniacal water and the impurities soluble in water will then settle, and may be readily drawn off.

Ammonia.—When the ammonia present in the water is sufficient in quantity to pay the cost and a profit upon its separation, it may be neutralized by the application of sulphuric or muriatic acid, and the solution evaporated to obtain sulphate, or muriate of ammonia, or it may be profitably employed in combination with other manures for a fertilizer. The carbonaceous matter that forms a stratum between the crude oils and the water is worthless unless it be used in the preparation of artificial fuel.

TREATMENT OF THE CRUDE COAL OILS.

The crude oils, being separated from their impurities, may at once be submitted to chemical treatment; but as a general rule, and especially when they are heavy and contain much tar, they should be first distilled. This distillation is made in a common iron still, protected from the action of the fire by fire brick, which equalizes the heat, consequently the expansion of the metal, and lessens the risk of fracture.

The "charge" of oil prepared as above, may be run into the still and distilled without the use of steam. But when it has been "run off" to four-fifths of the whole quantity, or when the part remaining in the still will be a thick pitch when cold, common steam should be gently let into the neck, or breast of the still. The steam immediately produces an outward current through the condensing apparatus and brings over all the remaining part of the oils, leaving a compact coke as the only residuum. Furthermore, it gradually diminishes the heat of the iron and prevents it from breaking. When the steam is thus let in, the fire is to be removed from beneath the still.

Continual Distillation.—In the first distillation of the crude oils, as they come from the retorts, and in subsequent ones, the oils may be slowly admitted into the still after it has become sufficiently heated and the oils begin to flow freely from the worm, or condenser. By the adjustment of a cock, a stream of the crude product may be permitted to flow through an iron tube into the still while it is in operation. The tube should dip beneath the oil in the still, the inflow of oil into which must not exceed the out-flow from the condenser. A greater amount of heat will be required for this operation than for the common method, as much of it is taken up by the cold oil constantly flowing inwards. By this mode a still working 1000 gallons may be made to run double that quantity without interruption, and steam may be applied in any manner before described.

The first distillate.—The first distillate of the crude oil should be separated into two parts, each of which requires somewhat different treatment. The first part is that which distils over from the commencement of the run until the oils in the receiver have a proof of 36° by hydrometer, or a specific gravity of 0·843.

These light hydro-carbons and the eupion they contain form the lamp oil. The quantity produced will depend upon the quality of the oil, or other material, whence they have been derived. This part of the distillate being pumped from the receiving tank, or otherwise removed, the remainder, or second part, is allowed to flow on until it assumes a greenish color at the end of the worm pipe, when steam, if not previously employed, may be let into the still and continued until the whole distillation is completed; the fire in the furnace beneath the still being withdrawn. A quantity of coke will be found to remain, amounting to ten or fifteen per cent. of the whole charge. This coke is excellent fuel, and after all its volatile matter has been expelled it may be employed in the clarification of sugar. When steam is not employed the residuum in the still must not be run down lower than a thick pitch. Coking in the still without steam is unsafe and hazardous to the iron.

The first part is then to be placed in an iron cistern and therein thoroughly agitated from one to two hours, with from four to ten per cent. of sulphuric acid, the object being to bring every particle of the impurities in contact with the acid. The quantity of acid to be used depends upon the character of the oils and the coal, heat, &c., employed in the retorts.

If too much acid is applied the oils will be partially charred and discolored; if too little, the impurities will not be oxidated, and the oils will change color. After the agitation of the oil and acid is completed, the mixture must remain at rest from six to eight hours, when the acid, with the chief part of the impurities, will have settled to the bottom of the vessel. They are then to be drawn off, and the remaining oil to be washed with ten or twenty per cent. of water. The water removes a part of the remaining acid, and carries off the soluble impurities. The acid now appears in the form of tar, and may afterwards be separated from the impurities for further use. After the water is withdrawn the charge is to be agitated two hours with from five to ten per cent., by measure, of a solution of caustic potash, or soda of specific gravity of 1·400. The hydrates of those alkalies may be used in the same manner; but the solution of caustic soda is generally preferred. Like the acid, the strength and quantity of the alkali must be varied according to the quality of the oils. After a repose of six hours or more, the alkali is to be withdrawn from the oil, and any further impurities rendered soluble, by its application, washed out with water. After the use of the water the oil should be perfectly neutral. When the water is withdrawn from it, it is to be run into a still for final rectification. Should any acid still remain in the charge, it may be distilled over two or four per cent. of the alkaline solution, or an equivalent quantity of lime, or soda ash with or without water. During the whole of these operations the oils and the several washes applied to them are to be kept at a temperature not lower than 90° Fah. This is conveniently done by means of steam coils fixed at the bottoms of the tanks in which the agitations are made. The agitator employed may be of any kind, if its action is efficient. Finally the oil is to be carefully distilled, with or without steam. A small quantity of the lightest product or eupion, which comes first from the condensing worm, is usually discolored, and may therefore be transferred to the succeeding charge.

The last distillation should be made slowly and with care, avoiding all fluctuations produced by an unsteady heat. If desired, the eupion may be taken off at the commencement of the distillation. It should be at proof 60°, or specific gravity, 0·733, or it may be allowed to run in with the lamp oil. When the distillate has reached proof 40°, or specific gravity 0·819, the remainder is to be transferred to the next charge, or the heavy oil, as being too dense for illuminating purposes.

The mixed oils intended for lamps have their disagreeable odor chiefly removed by allowing them to remain in flat open cisterns over weak solutions of the alkalies during a period of some days. Light also improves their color.

The alkalies employed in the foregoing treatment may be restored and used in subsequent purifications.

The oils of the second or heavy part of the first distillate are purified by the same means as described for the lighter oils, except that they require the application of more acid and stronger alkalies. All the oils distilled from them at proof 40° are to be added to the lamp oils. At the close of each distillation, and as the oils acquire greater density, the color grows darker and changeable, finally they are partially charred, and especially when they have been distilled without steam. These dark-colored oils may always be renovated by the use of acids and alkalies, the permanganates of potash and soda, and, finally, by distillation. The color of the lamp oils should not exceed a tinge of greenish yellow, when viewed in a clear glass flask six inches in diameter. If by accident, carelessness, or negligence, the oils treated by the foregoing method should be impure, they must be submitted to washing and redistillation.

PARAFFIN.

In general all the oils below 35° contain more or less paraffin; below 30° the paraffin is still more abundant. When the whole process has been well conducted those oils are to be placed in tanks in cool situations (thermometer at 40° and lower); the paraffin will then crystallize on the sides of the tanks in beautiful white silvery scales, from which the still liquid oils may be withdrawn.

MODE OF REFINING PARAFFIN.

The crude paraffin should be put into bags and submitted to pressure in a lard-oil press, or one which will gradually squeeze out the oil. It should then be pressed in a stearine press, cold, and all the oil remaining should be removed from it.

The paraffin is then melted in a wooden tank lined with lead, and furnished with a lead or iron steam coil in its bottom. One-half the weight of the paraffin of strong sulphuric acid is added and kept stirred up with it for four hours, and it is then permitted to settle for eight hours at a temperature of 190° Fah. The acids and impurities are then drawn off and the paraffin permitted to cool. It is then again pressed in the stearine press, and again melted in a steam heated vessel, in which it is agitated for three hours with a weak solution of caustic soda, when it is permitted to settle for six hours. The soda solution is then drawn off and the paraffin allowed to cool, when it is fit for moulding into candles upon being again melted.

The above process is effective when the crude paraffin does not contain an unusual amount of impurities. When it is very impure the treatment with acid and solution of soda must be repeated, using less of the acid for the second treatment. In this case, as in the purification of coal and petroleum oils, it is difficult to fix the exact quantity of the reagents to be used. Crude paraffin differs in quantity, some being quite white and pure, and other samples being black and heavy with tar and impurities.

The heavy oils, and those which drain from the paraffin, are excellent lubricators. They may be mixed with animal oils when it is desirable to give them greater consistence. As they do not readily oxidate when exposed to the air, they are peculiarly applicable as lubricants. The *gumming* complained of by machinists arises from the oxidation of the oils they have heretofore employed to relieve friction.

These heavy hydro-carbons, and even the solid paraffin itself, may be decarbonized and rendered suitable for lamps. Just in proportion to the amount, or number of the equivalents of carbon withdrawn from them, so are their boiling points lowered and their specific gravities diminished.

PURIFICATION OF PETROLEUM.

The usual method practised by refiners of petroleum is as follows:—

The petroleum is permitted to settle out water and impurities at a temperature of 90° F., and is then pumped into an iron still, such as have been described, when it is slowly and carefully distilled over open fire. Ten to twenty per cent. of the charge of crude oil is usually taken from it as naphtha. The remainder is distilled as low as possible, to give the greatest yield of lamp oil, which should not vaporize under 115° or 120° Fah. The residuum is made use of as mentioned in the second chapter. The lamp oil, which is usually 75 to 85 per cent. of the charge, is next well agitated with 1½ to 2 per cent. or 1½ to 2 gallons of sulphuric acid to the hundred gallons of the distillate, and permitted to settle for several hours. The acid and impurities are then drawn off. It is then agitated with water to remove the acid, and finally with three per cent. or three gallons to 100 gallons of the charge, of a solution of caustic alkali made as follows:—One gallon soda ash, half a gallon of fresh-slacked lime, five gallons of water, boiled twenty minutes in a kettle with steam jacket or over fire, and used at a specific gravity of 1·300 or 36° by Baumè's hydrometer. It is hardly possible to give the exact quantities of the reagents. The refiner may vary them to suit the petroleum to be treated.

It will be observed that the refining of petroleum is a very much cheaper process than the procuring of oil from coal. Distillation over carbonate of soda has been found advantageous in some cases. Distillation by superheated steam can be recommended for its causing the oils to distil over in a whiter and purer condition than when obtained by open fire. It is safe and perfectly under control, and is now used very largely in Great Britain for distilling shales and also for petroleum refining.

In distilling by superheated steam, it must be remembered that, although the heating pipes are at a temperature which would cause the steam, if confined, to burst them, it is not confined, but has a free passage into the still by means of the perforated pipe. Of course, if, by accident, the inlet valve became stopped, and the still exit closed at the same time, the superheater would be destroyed—but there is hardly a chance that either of those two things could occur.

In refining Canada petroleum a larger percentage of acid is used than in treating the Pennsylvania petroleum, but the process is, with that exception, the same. The odor of crude Canada petroleum is very offensive. This may be removed by adding nitric acid to the oil, in the proportion of one pint of acid to one hundred gallons of oil, and at the same time adding one gallon of chloride of lime. The oil must be well stirred and settled. This process removes the odor as far as the oil is concerned, when it is not being distilled. Nitric acid, or traces of it, would be very harmful in distillation, and the original odor would return. After the petroleum has been refined, the addition of the above quantities of nitric acid and chloride of lime, and thorough agitation, will for the greater part remove the odor of the refined oil.

Many attempts have been made to refine petroleum by filtration, or by agitation with various chemicals. A filter of bone-black will remove the color in some degree, but it would be too expensive a process on the large scale, even were it to render the oil of a uniform gravity of 819, or proof 40° Baumè—a thing which it cannot do. Oxalic ether exercises a bleaching effect upon petroleum, but it does not render it fit for lamps, and would be very expensive. It is doubtful whether any mode other than distillation will ever be found practicable in the purifying of petroleum oils. Those persons interested in petroleum should be very cautious in embarking in any new mode of refining petroleum. Nothing but the most absolute demonstration of the practicability of a new process could satisfy the author that anything except distillation can separate the naphtha, burning oil, and heavy oil, as his experiments in the direction of any other process, though they number several hundreds, have so far been productive of no practical result.

In refining petroleum or coal oils, care should be taken that the acid used be wholly removed by the alkali or water washing; many samples of petroleum oil are found to contain sulphuric acid in sufficient quantity to produce a most disagreeable and dangerous quantity of sulphurous acid gas in burning. This has been noticed by physicians upon visiting patients in the country, where the oil is most used. The atmosphere of the sick room is very soon made poisonous by the gas evolved by the night-lamp. Sulphurous acid gas is very irritating to the lungs and mucous membrane. The presence of sulphuric acid in the oil may be detected by adding a solution of chloride of barium to the oil, when a white precipitate will fall if any acid be present.

A piece of white blotting paper moistened with a solution of iodic acid and starch, held over the flame of the lamp, will become bluish-purple if sulphurous acid gas is being produced by the combustion. This test, however, may not always be correct, as iodine may be set free by other deoxidizing agents produced by combustion. The chloride of barium is, however, perfectly reliable.

PETROLEUM REFINERY.

The drawings on pages 162 and 163 represent a convenient arrangement of apparatus for refining petroleum. The superheated steam apparatus may be employed, or not, at the option of the refiner. If it is employed, the mode of operation would be as follows:—

The still, A, is filled to within six inches of the top to which the dome is bolted, by the pump, B, which draws the petroleum from the settler, E, in which is a coil of inch iron steam-pipe, to keep the temperature of the settler at 90° Fah. The settler is filled from the underground receiver, D, by the pump, B. This underground receiver is so arranged as to receive the petroleum from barrels being rolled over it and emptied.

The still being charged and the superheater having a fire under it which would give the steam a temperature of 300° Fah., the drip cock of the superheater (see cut on page 86) is opened, to permit any water or moist steam to escape, and afterwards the steam valve communicating with the superheater is opened gradually, and the steam permitted to pass into the still by the pipe, E, which is a two-inch wrought iron pipe bent under the dome, carried down the side of the still, coiled twice around the bottom of the still, and perforated with holes not less than one-eighth inch in

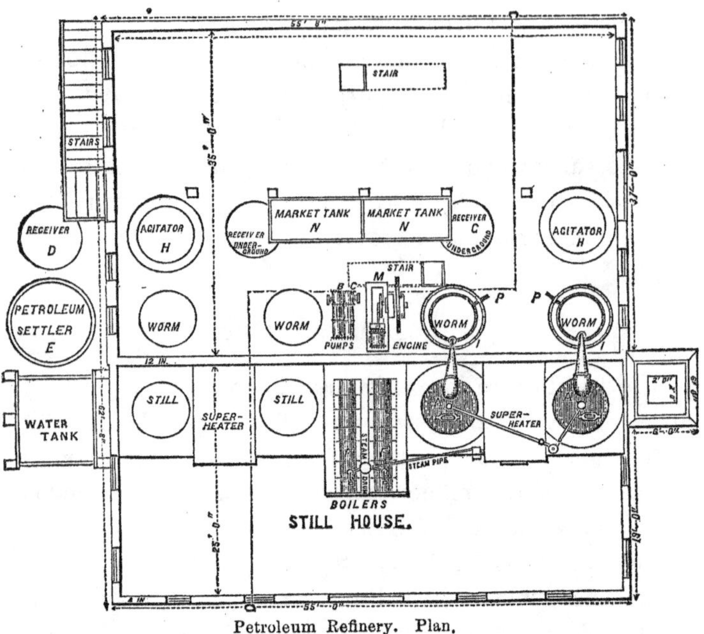

Petroleum Refinery. Plan.

diameter. As the hot steam enters the still there will be, for some time, a considerable condensation. This can be remedied by having a gentle fire under the still, but it is not absolutely necessary, as a still set in sand will soon become heated to the proper point. The worm tub, I, must be abundantly supplied with cold water from the water-tank outside the building, or by a pump. The inlet to the tub should be at the bottom and the overflow at the top, by which the hot water can be carried away to the drain, or to a vessel from which it can be pumped into the boilers. A great economy of heat can be effected by making use of hot instead of cold water for feeding the boilers. The distillate comes over about half water. As it proceeds, the heat under the superheater is increased, and more steam, at a pressure of 40 lbs. in the boiler, is

Elevation.

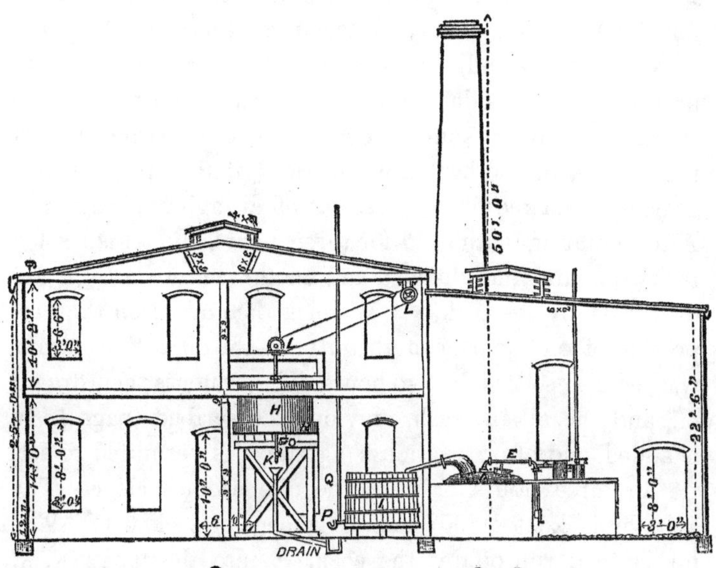

SECTION ON LINE C–D

passed into the still. At the close of the distillation, the superheater pipes will be at a temperature of 700° Fah., when the petroleum employed is of proof 40° to 45° Baumè. For heavier oils it will be at 800° Fah. The rapidity of the distillation is regulated by the valve on the superheater, and it can be made perfectly uniform with very little attention. There is no danger of "boiling over," as in distillation over open fire, which cannot always be controlled.

The naphtha which comes over first to the extent of about fifteen per cent. of the charge, is cut off at 58° Baumè generally, and permitted to run into a receiver. It may be redistilled, to purify it for market.

The remainder of the distillate is run into the underground receiver, G, whence it is pumped into the agitator, H, by the pump, C. It is allowed to settle in the agitator for three hours, and any water from it is run off by the cock, K. The acid, one and a half to two per cent., as desired, is then added to the oil in the agitator, and the stirring machinery put in motion by the gearing, L, and the engine, M. When atmospheric agitation is used, this gearing is not required. (See cut of air agitator, page 84.) After agitating with acid for three hours, the oil is settled for two hours, and the acid and tar thrown down by it are drawn off by the cock, K. Water is then added in the proportion of one quarter of the oil, and the oil is thoroughly washed and settled for two hours. The water is then drawn off, and the alkali wash, previously described (page 158), is added, and after two hours' agitation, is permitted to settle for three hours, or until the oil becomes clear. The agitator is kept at 90° Fah. by the steam-jacket, R. The oil is then run off by the cock, O, into the tanks, N, N, where it may be allowed to settle for market.

The heavy tar remaining in the still is drawn off by a cock placed near its bottom, after a few hours, or when it has cooled sufficiently to prevent its taking fire on exposure to the air.

In the drawings, the stills and superheaters are in sets each side of the boilers. This is to save room, and to have each set independent of the other, so that in case of repairs being needed to one set, the whole work would not be deranged and hindered. The building is capable of two more sets of stills and boilers on the opposite side, thus doubling its capacity. The agitator projects a convenient distance above the floor of the upper story to admit of easy handling of the carboys of acid which are to be emptied into it. This upper floor also affords space for making the alkali wash, which, when caustic soda is used, is simply by dissolving it in water until the proof is 36° by Baumè's lye meter. When soda ash is employed, the mode already described will give it sufficient causticity. Barrels, cans, and many other articles, can be stored on this floor.

Where superheated steam is not used, the boilers will not be required of the size shown in the drawing. The manipulation with the acid and alkali will be the same, however. In regular working, one of each set of stills should be kept always at work, the fire under the superheater never being permitted to become very low, or so low as to cause cracking of the pipes. There is no economy of fuel in permitting the fire to die out.

The syphon upon the tail-pipe, P, is useful in distilling Canada petroleum, or the tar of stearine distillation, as it forms a trap, which, connecting with the pipe, Q, carries the offensive gas generated to the outer air.

The pumps should be capable of pumping the whole

charge in one hour; and the pump used for pumping crude petroleum should never be used for refined oil. When the agitator becomes dirty, or after each settling of acid or alkali, a steam jet from a rubber hose may be used against its sides with good effect. Steam at forty lbs. is an excellent scrubber, and is very convenient, as it will reach corners and crevices very readily.

Care must be taken to keep all vessels used in oil making perfectly clean. A jet of steam carried into the throat of the gooseneck, by a pipe, as in the drawings on pages 148, 149, is very useful in cooling down after the charge is off. Such a refinery as the one just described can be erected at the following cost:—

		IN GOLD.
4 Cast-iron Stills, 7 feet diameter, 4 feet deep, contents 1145 gallons, working contents 900 gallons, ...	$500	$2,000
4 Condensing Worms, 100 feet long, tapering from 4 inches to $2\frac{1}{2}$ inches, in tanks complete, ...	200	800
2 Superheaters, each 60 feet of 2-inch pipe, in furnace, complete, . . .	350	750
2 Boilers, 36 inch shell, 2 12-inch flues, 16 feet in length, ...	400	800
2 Washers or Agitators, 7 feet diameter, 5 feet deep, ...	140	280
1 Petroleum Settler, 10 feet diameter, 8 feet deep, ...		200
2 Underground Receivers, 7 feet diameter, 5 feet deep, ...	100	200
1 Underground Receiver, 8 feet diameter, $6\frac{1}{2}$ feet deep, ...	100	100
1 5-Horse Engine, ...		300
2 Steam Pumps (15 gallons per minute),	250	500
Gearing for washers, ...		200
Pipes, cocks, and fittings, . . .		500

Building 62 feet 8 inches long, 85 feet wide, 2 stories, with shaft 2 feet by 2; flue, 50 feet high, complete, . . .	4,000
Setting boilers and stills, . . .	600
	$11,230

It is capable of refining 2,000 gallons of petroleum per day, and would turn out about 1,500 gallons of refined oil in that time. It would require one superintendent, two engineers, four still men, and four helpers.

The cost of many of the petroleum refineries has been less than the sum named, and, of a few of them, ten times greater.

In many places the petroleum refinery consists of a common boiler still and worm, and a few vats. There are large works, however, in which the details are well carried out. The works of the Liverpool Oil Refining Company, at Bootle, near Liverpool, under the direction of Andrew McLean, are very complete and efficient. Those of the Kerosene Oil Company, on Newtown Creek, Long Island, New York, and those of Samuel Downer, at Corry, Pennsylvania, are also very extensive and perfect.

In some refineries, when air is used as an agitator, the pipes communicating with the air-pump or blower, pass near the boiler or furnace flues, and in this way heat the air before it is driven into the oil. It is supposed that hot air is a better oxidator than cold.

REFUSE OF OIL REFINERIES.

Formerly the acids and alkalies used in coal oil and petroleum refining were considered as waste materials, but they are now made use of in various ways.

The acid "bottoms," so called, are used in the manufacture of superphosphate of lime, the oil being partly removed.

In large works it would be an economy to recover the alkali by evaporation and calcination. This was successfully done by James Campbell, of Dayton, Ohio, while engaged in manufacturing coal oil at Charleston, Kanawha county, Western Virginia, some years ago.

In coal oil works there is a large quantity of illuminating gas generated. This will hereafter be put to use, no doubt, when the situation of the works will admit of it.

At situations where coal, or the supply of coke, is insufficient, the gas may be most advantageously employed for producing steam, and for all the distillations required in making and purifying the oils. For those purposes it is superior to any other kind of fuel, as the heat may be increased or diminished instantaneously at the will of the operator. For heating, the gas requires no purification, and recent improvements in producing heat by this agent will supply the highest temperature required.

Coke.—When the coal employed affords a good coke it is used for fuel; the coke of Boghead coal and the bituminous shales is of little value. Some of the asphaltums and bitumens afford a small residue of fuel.

Ashes.—Ashes collect around oil manufactories in large quantities, and they differ in their composition according to the nature of the coal consumed. In all cases where they contain any considerable percentage of lime, they will be found valuable fertilizing agents for certain soils.

Ammoniacal water.—Whenever nitrogen enters into the composition of the coal, shale, or other material distilled in the retorts, ammoniacal water will be one of the products, and upon it the lighter oils will repose in the receiving-vessels. The quantity of ammonia is often very considerable.

Sulphate of ammonia.—To prepare the sulphate of ammonia from the crude ammoniacal water, the latter is to be saturated with sulphuric acid, and evaporated in a cast-iron boiler. The saturation may be made in a leaden vessel, and the evaporation performed by steam. When the liquid has attained a specific gravity of 1·400, or thereabouts, it should be run into a vessel lined with lead, and crystallized. Another mode consists of distilling the ammoniacal water, and conducting the distillate into a solution of sulphuric acid of spec. grav. 1·700. In this case the sulphate of ammonia is precipitated, and may be dipped out with ladles.

Chlorohydride of ammonia (*sal ammoniac*).—To form the sal ammoniac of commerce, the ammonial water is to be saturated with hydrochloric acid (muriatic acid). It is usually evaporated in vessels of lead, and then run into wooden coolers. The salt is then to be dried in stoves, and finally sublimed in iron pots with large domes. Some days are required to complete the last operation.

The pitch resulting from distillation of coal oil or petroleum is now used for varnish, roofing, and other purposes.

It will always be worth the attention of those in the trade to recover, in some way, a part at least of the value of the chemicals employed. Nothing can be considered as absolutely a waste product which has been formed in the various manipulations of the oil refiner.

The acid bottoms, or the tarry matter thrown down by the acid in refining petroleum, is now said to be successfully applied to the production of aniline. This may, in some degree, account for the very great fall in the price of aniline, but it is probably due to the increasing quantity manufactured abroad. The various aniline dyes, in crystals, can be now purchased for from $7 to $8 per lb. Two years ago the price was $200 for the same quantity.

BAUMÈ'S HYDROMETER.

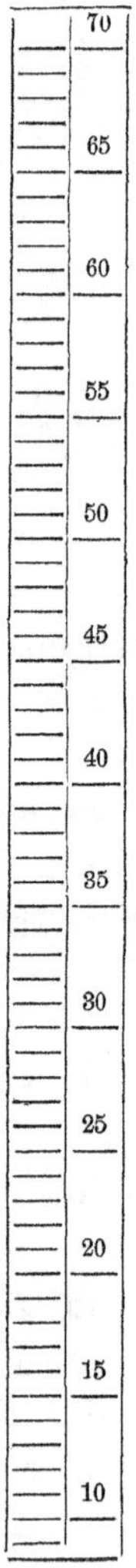

TABLE OF SPECIFIC GRAVITIES OF OILS,

CORRESPONDING TO DEGREES OF HYDROMETER.

SCALE.

70 65 60 55 50 45 40 35 30 25 20 15 10

Degrees of Hydrometer. Ther. 60° Fah.	Specific Gravity.	Degrees of Hydrometer. Therm. 60° Fah.	Specific Gravity.	Degrees of Hydrometer. Therm. 60° Fah.	Specific Gravity.
70	0 696	49	0·778	28	0·880
69	0·700	48	0·782	27	0·886
68	0·704	47	0·787	26	0·890
67	0·707	46	0·791	25	0·898
66	0·711	45	0·795	24	0·903
65	0·713	44	0·800	23	0·910
64	0·718	43	0·804	22	0·915
63	0·722	42	0·808	21	0·921
62	0·725	41	0 813	20	0·927
61	0·729	40	0·819	19	0·933
60	0·733	39	0·824	18	0 944
59	0·737	38	0·828	17	0·940
58	0·741	37	0·833	16	0·951
57	0·745	36	0·838	15	0·959
56	0·749	35	0·843	14	0·966
55	0·753	34	0·848	13	0·971
54	0·757	33	0·854	12	0·979
53	0·761	32	0·860	11	0·986
52	0·765	31	0·864	10	0·994
51	0·769	30	0·869		
50	0 773	29	0·875		

The scale in common use for ascertaining the specific gravity of fluids, lighter or heavier than water, is that of Baumè. It is rarely made with sufficient care to insure accurate correspondence between the degrees marked upon it and the true specific gravity. It is, however, accurate enough for general purposes, and is the hydrometer referred to in this work, accidentally omitted to be mentioned in its proper place.

Baumè's hydrometer for liquids heavier than water, such as is used for acids, solutions of salts, etc., is the one referred to when the strength of alkaline solutions is mentioned, unless any other is indicated by name.

PYROMETER FOR COAL AND PETROLEUM OILS.

This instrument is used for determining the exact temperature at which an oil will inflame. Instead of being sold by proof by the hydrometer, the oil is tested and rated, as it will stand 110° Fah. or 115° Fah. without taking fire. The cuts below are representations of the pyrometer made by Giuseppe Tagliabue, of New York.

FIG. 1 is a perspective view of the PYROMETER, as it appears when prepared for testing the temperature of the oil at the moment of the explosion of its vapor. FIG. 2 represents the instrument when prepared for measuring the inflaming point of the liquid oil.

The bath, *B*, is supported in its cylindical stand, *C*, made of metal, which has an aperture near the bottom, to admit of the insertion of a small spirit lamp; when this lamp is lighted, the oil is of course gradually heated by the water, and it emits vapor with a rapidity proportioned to its volatility; this vapor is mingled with atmospheric air, which enters through two perforations, *d d*, (FIG. 1), in the cover

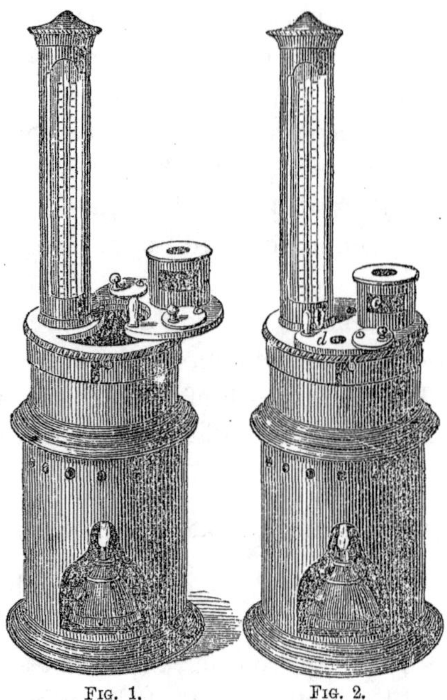

FIG. 1. FIG. 2.

of the cup, *A*, and thus forms an explosive mixture that ascends into the cylinder, *F*. On the insertion of a lighted taper in the aperture, *e*, in the cylinder, the lowest explosive temperature of the oil is accurately indicated by a slight explosion or "puff," and a simultaneous inspection of the mercury in the thermometer. After partly removing the cover of the lamp by revolving it, and holding the flame in actual contact with the escaping vapor until the oil burns, the thermometer will precisely indicate the degree of inflammability.

CEMENT FOR IRON FLANGED OR SOCKET JOINTS.

Fine iron borings or filings, 5 lbs.; sal ammoniac, in powder, 2 oz.; water to mix to a paste, which is well rammed into the joint by blunt chisels to suit. The filings should be as fine as possible. In flanged joints the cement is prevented from entering the pipe by a thin ring of iron placed inside the line of the bolt-holes. The bolts should be screwed up after the joint is made with the cement.

A very excellent luting for joints which are to be often broken, such as manhole plates, is made of fine slacked lime and common glue. The glue should be dissolved to a thick jelly, and the lime added, until a stiff paste is formed. No more should be mixed than wanted for immediate use, which may apply also to the iron cements. In making rust-joints, as the iron cementing is called, the flanges, or sockets, should be cleaned with a solution of muriatic acid before the joint is rammed. A good fine clay, free from sand or grit, is the cheapest luting for retort lids.

COST OF ARTIFICIAL LIGHT.

Illuminating Material.	Cost per lb. or gallon.	Consumption in four hours.	Cost of Light per hour.
Wax candles (red)	$0 50 per lb.	532 grains	1·068 cts.
" " (green)		458 "	
Paraffin candles, 6's.	0 60 "	567 "	1·395 "
Tallow " 6's.	0 15 "	563 "	0·324 "
Sperm " 4's.	0 40 "	587 "	0·984 "
Star "	0 25 "	636 "	0·688 "
Lard oil	1 20 per gallon.	12·61 ozs. fluid	2·096 "
Burning fluid . . .	0 75 "	5·09 "	0·746 "
Kerosene	1 20 "	3·89 "	0·912 "
ADDENDA.			
Petroleum oil. . .	1 00 "	3·24 "	0·860 "
New York coal gas .	2 50 per 1000 ft.	4 feet burner.	1·000 "

The foregoing table, relating to the cost of Artificial Light, has been extracted from the statement of Dr. Charles M. Wetherill, and published in *The American Gas-Light Journal*, May 1, 1860.

LOCALITY, DEPTHS, AND YIELD OF SOME OF THE PRINCIPAL PETROLEUM WELLS OF THE UNITED STATES.

The "Burned" well, on Oil Creek, Pennsylvania, was completed in April, 1861, at a depth of 330 feet. On the afternoon of April 17, while the workmen were engaged in tubing, a stream of gas suddenly lifted the tools out of the well and leaped above the derrick in a continuous and sickening volume. The engineer put out his fires, and then, with the rest of the hands, fled from the sickening odor that oppressed the air. A crowd collected, some one in which, approaching too near, suddenly ignited the gas, which went off with a terrific explosion, setting fire, of course, to the stream of oil issuing from the well. The conflagration that ensued, and which continued for four days and nights, finally destroyed the well. The lives of several persons were lost. The well has not yielded any since.

The "Brawley" well, at a depth of 503 feet, began to flow in the summer of 1861, yielding 600 barrels per day. After flowing a year and a half, the yield began to diminish. It speedily ran down to nothing.

The "Van Slyke" well "struck oil" in the fall of 1861, at a depth of about 500 feet, and first flowed at the rate of 600 barrels per day It also gave out in about a year and a half.

The "Big Phillips" well struck oil in October, 1861, at a depth of 480 feet. The estimated quantity of the original flow was from 3,000 to 4,000 barrels per day. The rush of oil was so overwhelming, that it was several days before the well could be tubed; 40,000 to 50,000 barrels of oil were lost in the creek before the workmen finally got control. The well was subsequently (like every other well yielding at that period) not permitted to flow under anything like full headway, the price of oil being so low as not to pay. The flow began to decrease about the latter part of 1862. In this year another well, the "Woodford," was put down near, which tapped the same vein of oil, and assisted in diminishing the flow. The "Big Phillips" is now running at the rate of 325 barrels per day. It is believed to be the only well which began flowing without having been previously tubed.

The "Woodford" well, alluded to above, was originally a 1,500 barrel. Its yield began to decrease in 1863, and finally ceased. Being resuscitated, it is now pumping 50 barrels per day.

The "Jones" well, put down in the latter part of 1862, within 30 feet of the "Woodford," tapped the same vein, flowing 400 barrels per day. Its flow decreased gradually until the well had to be pumped. It is now doing nothing.

The "Noble" well struck oil in April, 1863. Its maximum daily yield was between 1,900 and 2,000 barrels. It flowed six months with undiminished volume, when it began to decrease. It was flowing until the 1st of February, 1865, at the rate of 150 to 200 barrels per day, when an accident stopped it. This well is said to have netted its owners over $3,000,000.

The "Empire" well was sunk in the fall of 1861, and began flowing from 2,500 to 3,000 barrels per day. The flow continued diminishing gradually for something over two years, when it stopped. The well lay idle about a year. In the summer of 1864, an air-pump was applied, which caused the well to resume flowing lightly—five or six barrels per day. The flow then slowly increased to 140 barrels. The well is now yielding 110 barrels per day.

The "McKinley" flowing well, on Oil Creek, is remarkable for the permanence of its yield. It has given from 50 to 60 barrels per day for nearly three years. It is under very excellent management.

The "Williams and Stanton" wells, three in number, yield 150 barrels per day in all. These wells must have pierced the same reservoir, as it is found that by stopping two of them, the oil will flow from the third at the above rate.

The "Reed" well, on Cherry Run, gives 280 barrels per day.

The "Fox" well, at Petroleum Centre, yields 160 barrels per day.

In Western Virginia, the "Burning Spring" well, the "Llewellyn," and others have been noted for their yield of oil.

It has been computed, that in the beginning of the present year, the district of Oil Creek—the most productive of the Pennsylvania oil regions, and embracing an area of 3,200 acres—contained 480 wells already sunk, 542 wells in progress, and 189 producing wells. The average daily yield from this district was estimated at 4,000 barrels.

INDEX.

www.ingramcontent.com/pod-product-compliance
Lightning Source LLC
LaVergne TN
LVHW011224110826
845150LV00006B/1543

* 9 7 8 1 4 2 5 5 1 6 2 7 7 *